New Studies in Biolog

Photosynthesis

Fourth edition

D. O. Hall
Ph.D.
Professor of Biology,
King's College, University of London

K. K. Rao
Ph.D
Honorary Lecturer in Biology
King's College, University of London

The right of the
University of Cambridge
to print and sell
all manner of books
was granted by
Henry VIII in 1534.
The University has printed
and published continuously
since 1584.

CAMBRIDGE UNIVERSITY PRESS
Cambridge
New York   Port Chester
Melbourne   Sydney

Published by the Press Syndicate of the University of Cambridge
The Pitt Building, Trumpington Street, Cambridge CB2 1RP
40 West 20th Street, New York, NY 10011–4211, USA
10 Stamford Road, Oakleigh, Victoria 3166, Australia

First published by Edward Arnold 1972
Second edition 1977
Third edition 1981
Fourth edition first published by Edward Arnold 1987 and first
published by Cambridge University Press 1992

Printed in Great Britain at the University Press, Cambridge

British Library cataloguing in publication data
Hall, D. O.
    Photosynthesis.—4th ed.—(The Institute
    of Biology's studies in biology,
    ISSN–0537–9024)
    1. Photosynthesis
    I. Title   II. Rao, K. K.   III. Series
    581.1'3342   QK882

Library of Congress cataloguing in publication data available

ISBN 0 521 42806 8 paperback

General Preface to the Series

Recent advances in biology have made it increasingly difficult for both students and teachers to keep abreast of all the new developments in so wide-ranging a subject. The New Studies in Biology, originating from an initiative of the Institute of Biology, are published to facilitate resolution of this problem. Each text provides a synthesis of a field and gives the reader an authoritative overview of the subject without unnecessary detail.

The Studies series originated 20 years ago but its vigour has been maintained by the regular production of new editions and the introduction of additional titles as new themes become clearly identified. It is appropriate for the New Studies in their refined format to appear at a time when the public at large has become conscious of the beneficial applications of knowledge from the whole spectrum from molecular to environmental biology. The new series is set to provide as great a boon to the new generation of students as the original series did to their parents.

1986

Institute of Biology
20 Queensberry Place
London SW7 2DZ

Preface to the Fourth Edition

In this expanded fourth edition we have retained the inherent features of the previous editions which seem to appeal to students and teachers alike, viz. basic descriptive ideas on the process of photosynthesis, an historical outline as to how these ideas developed, the current status in our understanding of photosynthesis, and an overview as to where modern research is leading. At the same time we have taken this opportunity to revise some of the previously accepted concepts in the light of recent discoveries and to expand the book by the inclusion of topics which are allied to photosynthetic energy conversion and chloroplast genetics. The sections on chloroplast structure and photosynthetic electron transport have been updated with the inclusion of new figures and a table showing the properties of recently identified polypeptides and redox species along with components already accepted as intermediates of the electron transfer chain. The distribution of the photosystem complexes in the membrane and the molecular properties and relative orientation in the membrane of various polypeptides are also illustrated by new figures. The chapter on CO_2 fixation is expanded to include sucrose and starch synthesis, nitrogen and sulphur metabolism, fatty acid biosynthesis, and oxygen exchange.

A major breakthrough in photosynthesis research was the recent crystallization of a bacterial photoactive reaction centre complex and determination of its structure by X-ray analysis. This structure is outlined in the chapter on bacterial photosynthesis which now contains a comparison of plant and bacterial photosynthesis and evolution of photosynthesis as additional topics.

The section dealing with research on photosynthesis has been thoroughly revised and extended to bring to focus areas such as chloroplast DNA structure, photoinhibition, and energy transfer across the two photosystems. As usual the reading list has been revised and updated.

We are very grateful to colleagues who kindly provided us with figures, some hitherto unpublished. The suggestions from readers and reviewers of the previous editions have been most welcome and we look forward to further comments on the contents of this new edition.

London, 1986   D. O. H.
                                                                      K. K. R.

Contents

1

Importance and Role of Photosynthesis

1.1 Ultimate energy source

Hardly a day goes by without the importance of photosynthesis being brought to our attention. All our food and our fossil and biological fuels (biomass) are derived from the process of photosynthesis. Increasingly the products of photosynthesis are being sought to feed and fuel the world and also to provide chemicals and fibres. Thus an understanding of the fundamental and applied aspects of photosynthesis is now essential to a wide range of scientists and technologists – from agriculture and forestry through ecology and biology to chemistry, genetics and engineering. It is this universality that attracts varied approaches to studying photosynthesis and makes it such an exciting field of work to so many different types of people. We hope that this becomes evident in our book.

The term photosynthesis literally means building up or assembly by light. As used commonly, photosynthesis is the process by which plants synthesize organic compounds from inorganic raw materials in the presence of sunlight. All forms of life in this universe require energy for growth and maintenance. Algae, higher plants and certain types of bacteria capture this energy directly from the solar radiation and utilize the energy for the synthesis of essential food materials. Animals cannot use sunlight directly as a source of energy; they obtain the energy by eating plants or by eating other animals which have eaten plants. Thus the ultimate source of all metabolic energy in our planet is the sun and photosynthesis is essential for maintaining all forms of life on earth.

We use coal, natural gas, petroleum, etc. as fuels. All these fuels are decomposition products of land and marine plants or animals and the energy stored in these materials was captured from the solar radiation millions of years ago. Solar radiation is also responsible for the formation of wind and rain and hence the energy from windmills and hydro-electric power stations could also be traced back to the sun.

The major chemical pathway in photosynthesis is the conversion of carbon dioxide and water to carbohydrates and oxygen. The reaction can be represented by the equation:

$$CO_2 + H_2O \xrightarrow[\text{plants}]{\text{sunlight}} \underset{\text{Carbohydrate}}{[CH_2O]} + O_2$$

The carbohydrates formed possess more energy than the starting materials, namely CO_2 and H_2O. By the input of the sun's energy the energy-poor compounds, CO_2 and H_2O, are converted to the energy-rich compounds, carbohydrates and O_2. The energy levels of the various reactions which lead up to the overall equation above can be expressed on an oxidation-reduction scale ('redox potential' given in volts) which tells us the energy available in any given reaction – this will be discussed later in Chapter 4. Photosynthesis can thus be regarded as a process of converting radiant energy of the sun to chemical energy of plant tissues.

1.2 The carbon dioxide cycle

The CO_2 content of the atmosphere remains almost constant in spite of its depletion during photosynthesis (however, see §8.13, P.108). All plants and animals carry out the process of respiration (in mitochondria) whereby oxygen is taken from the atmosphere by living tissues to convert carbohydrates and other tissue constituents eventually to carbon dioxide and water, with the simultaneous liberation of energy. The energy is stored in ATP (adenosine triphosphate) and is utilized for the normal functions of the organism. Respiration thus causes a decrease in the organic matter and oxygen content and an increase in the CO_2 content of the planet. Respiration by living organisms and combustion of carbonaceous fuels consumes on average about 10 000 tonnes of O_2 every second on the surface of the earth. At this rate all the oxygen of the atmosphere would have been used up in about 3000 years. Fortunately for us the loss of organic matter and atmospheric oxygen during respiration is counter-balanced by the production of carbohydrates and oxygen during photosynthesis. Under ideal conditions the rate of photosynthesis in the green parts of plants is about 30 times as much as the rate of respiration in the same tissues. Thus photosynthesis is very important in regulating the O_2 and CO_2 content of the earth's atmosphere. The cycle of operations can be represented as shown in Fig. 1.1. All the CO_2 in the atmosphere is cycled through plants, via photosynthesis, every 300 years and all the O_2 is cycled every 2000 years.

It should be made clear that the energy liberated during respiration is finally dissipated from the living organism as heat and is not available for recycling. Thus for millions of years, energy has been constantly removed from the sun and wasted as heat in the earth's atmosphere. But fortunately there is still enough energy available from the sun for photosynthesis to continue for many hundreds of millions of years.

1.3 Efficiency and turnover

Photosynthetic efficiency on a global basis may be defined as the fraction of photosynthetically active radiation (PAR) that falls on the earth's surface that is converted to *stored* energy by photosynthesis in the biosphere.

The solar energy striking the earth's atmosphere every year is equivalent to

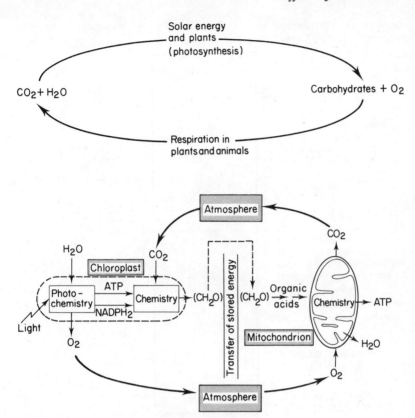

Fig. 1.1 The CO_2 and O_2 cycle in the atmosphere and the cell.

about 56×10^{23} joules (J) of heat. Of this roughly half is reflected back by the clouds and by the gases in the upper atmosphere. Of the remaining radiation that reaches the earth's surface only 50% is in the spectral region of light that could bring about photosynthesis, the other half being weak infrared radiation (Fig. 1.2). Thus the annual influx of energy of photosynthetically active radiation, i.e. from violet to red light, to the earth's surface is equivalent to about 15×10^{23} J. However, some 40% of this is reflected by ocean surface, deserts, etc. and only the rest can be absorbed by the plant life on land and sea. Recent estimates of the total annual amount of biomass (plant matter produced by photosynthesis) are about 2×10^{11} tonnes of organic matter which is equivalent to about 3×10^{21} J of energy. About 40% of this organic matter is synthesized by phytoplankton, minute plants living near the surface of the oceans. The annual food intake by the earth's human population (assuming the population to be 4800 million) is approximately 900 million tonnes or 14.5×10^{18} J. Thus the average coefficient of utilization of the incident photosynthetically active radiation by the entire flora of the earth is only about 0.2% ($3 \times 10^{21}/15 \times 10^{23}$)

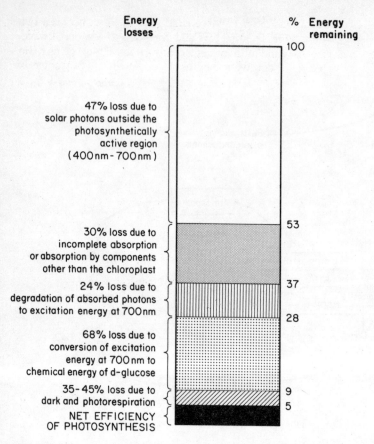

Fig. 1.2   Summary of the energy losses in photosynthesis as sunlight falls on a leaf at 25°C (from Bolton and Hall, 1979).

and of this less than 0.5% (14.5 × 10^{18}/3 × 10^{21}) is consumed as nutrient energy by mankind. It is interesting that the total consumption of energy by the world including biomass in 1984 was 3.6 × 10^{20} J – this was only one-tenth of the energy stored by photosynthesis! In fact, the energy content of the biomass standing on the earth's surface today (90% trees) is equivalent to all our proven reserves of fossil fuel, i.e. oil, gas and coal; also the total resources of fossil fuel stored below the earth's surface only represents about 100 years of net photosynthesis.

1.4 Spectra

Light is a form of electromagnetic radiation. All electromagnetic radiation has wave characteristics and travels at the same speed of 3 × 10^8 m s^{-1} (*c*, the speed of light). But the radiations differ in wavelength, the distance between two successive peaks of the wave. Gamma rays and X-rays have very small wavelengths (less than one thousand millionth of a centimetre, 10^{-11} m) while radio waves are in the order of 10^4 cm. Wave lengths of visible light are conveniently expressed by a unit called a nanometre. One nanometre is one thousand millionth of a metre (1 nm = 10^{-9} m). It has been known since the time of Isaac Newton that white light can be separated into a spectrum, resembling the rainbow, by passing light through a prism. The visible portion of this spectrum ranges from the violet at about 380 nm to the red at 750 nm (Fig. 1.3).

The atmosphere of the sun consists mainly of hydrogen. The energy of the sun is derived from the fusion of four hydrogen nuclei to form a helium nucleus. The fusion process is a multi-step reaction which can be simplified as $4H \rightarrow He + h\nu$ (energy). The mass of the He nucleus is less than the total mass of 4H; the mass lost during fusion is converted to energy. The energy liberated during the nuclear fusion maintains the surface temperature of the sun around 6000 K. The sun radiates energy representing the entire electromagnetic spectrum but the earth's atmosphere is transparent only to part of the infra-red and ultraviolet light and all the visible light. The ultraviolet waves which are somewhat shorter than the shortest visible light waves are absorbed by the oxygen and ozone of the upper atmosphere. This is fortunate since ultraviolet radiations are harmful to living organisms. At 6000 K, the temperature of the sun, the maximum intensity of emitted light lies in the orange part of the visible spectrum, around 600 nm.

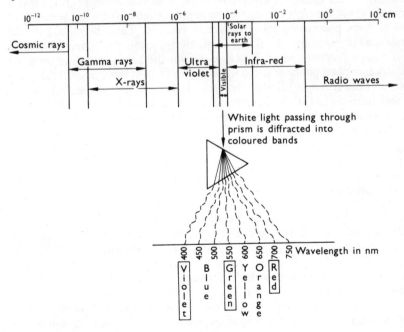

Fig. 1.3 Spectra of electromagnetic radiation.

1.5   Quantum theory

In 1900 Max Planck enunciated the theory that the transfer of radiation energy within a hot object involved discrete 'units' of energy called quanta. Planck's quantum theory can be expressed mathematically as $E = h\nu$ where E is the energy of a single quantum of radiation, ν is the frequency of the radiation (frequency is the number of waves transmitted in unit time), and h a constant. The Planck's constant (h) has the dimensions of the product of energy and time and its value in the c.g.s. system is 6.626×10^{-34} J s. Planck's theory proposes that an oscillator of fundamental frequency (ν) would take up energy $h\nu$, $2h\nu$, $3h\nu \rightarrow nh\nu$, but it could not acquire less than a whole number of energy quanta. Five years later Albert Einstein extended Planck's theory to light and proposed that light energy is transmitted not in a continuous stream but only in individual units or quanta. The energy of a single quantum of light or *photon* is the product of the frequency of light and Planck's constant, i.e. $E = h\nu$. Since frequency is inversely related to wavelength, it follows that photons of short wave light are more energetic than photons of light of longer wavelength, i.e. at one end of the spectrum photons of blue light are more energetic than those of red light at the other end.

For photosynthesis to take place the pigments present in plant tissues should absorb the energy of a photon at a characteristic wavelength and then utilize this energy to initiate a chain of photosynthetic chemical events. We will learn later that an electron is ejected from the pigment immediately after the absorption of a suitable quantum of light. It should be emphasized that a photon cannot transfer its energy to two or more electrons nor can the energy of two or more photons combine to eject an electron. Thus the photon should possess a critical energy to excite a single electron from the pigment molecule and initiate photosynthesis. This accounts for the low efficiency of infra-red radiation in plant photosynthesis since there is insufficient energy in the quantum of infra-red light. Certain bacteria, however, contain pigments which absorb infra-red radiation and carry out a type of photosynthesis which is quite different from plant-type photosynthesis in that no O_2 is evolved during the process (see Chapter 7).

1.6   Energy units

According to Einstein's law of photochemical equivalence a single molecule will react only after it has absorbed the energy of one photon ($h\nu$). Hence one mole (gram-molecule) of a compound must absorb the energy of N photons ($N = 6.023 \times 10^{23}$, the Avogadro number), i.e. $Nh\nu$, to start a reaction. The total energy of photons absorbed by one mole of a compound is called an Einstein, i.e. one Einstein = 6.023×10^{23} quanta.

Let us calculate the energy of a mole (or Einstein) of red light of wavelength 650 nm (6.5×10^{-7} m). The frequency, $\nu = c/\lambda$ = speed of light/wavelength of light

$$\nu = 3.0 \times 10^8/6.5 \times 10^{-7} = 4.61 \times 10^{14}$$

$E = Nh\nu,$
i.e. Energy = number of molecules $\times$ Planck's constant $\times$ frequency

$$\therefore E = 6.023 \times 10^{23} \times 6.626 \times 10^{-34} \times 4.61 \times 10^{14}$$

$$= 18.40 \times 10^4 \text{ joules} = \text{energy of one Einstein of red light}$$

$$\text{or } E = 18.40 \times 10^4/4.184 \times 10^3 = 43.98 \text{ kcal}$$

(One kilocalorie, kcal, is equal to 4.184×10^3 joules.) Thus 1 mole of red light at 650 nm contains 18.40×10^4 joules of energy.

The energy of photons can also be expressed in terms of electron volts. An electron volt, eV, is the energy acquired by an electron when it falls through a potential of 1 volt, which is equal to 1.6×10^{-19} joules. If 1 molecule of a substance acquires an average energy of 1 eV the total energy of a mole can be calculated to be 9.64×10^4 joules ($1.6 \times 10^{-19} \times 6.023 \times 10^{23}$). Thus the energy of 1 mole of 650 nm light is equal to 1.91 eV ($18.40 \times 10^4/9.64 \times 10^4$).

Table 1.1 Energy levels of visible light.

Wavelength	Colour	Joules per mole	kcal per mole	Electron volts per photon
700 nm	Red	17.10×10^4	40.87	1.77
650 nm	Orange-red	18.40×10^4	43.98	1.91
600 nm	Yellow	19.95×10^4	47.68	2.07
500 nm	Blue	23.95×10^4	57.24	2.48
400 nm	Violet	29.93×10^4	71.53	3.10

1.7 Measurement of photosynthetic irradiance

Historically light intensity was measured in terms of lumens, a lumen being defined as the luminous flux on a unit surface all points of which are at unit distance from a uniform point source of one candle. Intensity of illumination was expressed either as foot candles (1 lumen per sq ft) or lux (one lumen per sq metre). Nowadays photobiologists prefer to measure light energy incident on a surface, i.e. radiant flux density or *irradiance*, in terms of the units of power as Watts per sq metre (Wm^{-2}). Since photochemical reactions in photosynthesis depend more on the *number* of photons incident on a surface rather than on the energy content of these photons, it is more logical to express photosynthetic irradiance in terms of the number of quanta (photons) falling on unit surface in unit time, i.e. as the *photon flux density*. The photon (or quantum) flux density (Q) in a particular wavelength region is measured in units of mol m^{-2} s^{-1} where a mol is 6.023×10^{23} (Avogadro's number) quanta or photons. Since an Einstein (E) is defined as 6.023×10^{23} quanta, Q can also be expressed as E m^{-2} s^{-1}. A more practical unit to express photosynthetic photon flux density

(PPFD) is μmol m^{-2} s^{-1} or μE m^{-2} s^{-1}. For example, the solar irradiance reaching the earth's surface in full sunlight is approximately 950 Wm^{-2} or 95 000 lux in which the photon flux density of photosynthetically active radiation (400–700 nm) is 1800 μmol m^{-2} s^{-1}.

2

History and Progress of Ideas

2.1 Early discoveries

In the early half of the seventeenth century the Flemish physician van Helmont grew a willow tree in a bucket of soil feeding the soil with rain water only. He observed that after five years the tree had grown to a considerable size though the amount of soil in the bucket had not diminished significantly. Van Helmont naturally concluded that the material of the tree came from the *water* used to wet the soil. In 1727 the English botanist Stephen Hales published a book in which he observed that plants used mainly *air* as the nutrient during their growth. Between 1771 and 1777 the famous English chemist Joseph Priestley (who was one of the discoverers of oxygen) conducted a series of experiments on combustion and respiration and came to the conclusion that green plants were able to reverse the respiratory processes of animals. Priestley burnt a candle in an enclosed volume of air and showed that the resultant air could no longer support burning. A mouse kept in the residual air died. A green branch of mint, however, continued to live in the residual air for weeks. At the end of this time Priestley found that a candle could burn in the reactivated air and a mouse could breathe in it. We now know that the burning candle used up the *oxygen* of the enclosed air which was replenished by the photosynthesis of the green mint. A few years later the Dutch physician, Jan Ingenhousz, discovered that plants evolved oxygen only in *sunlight* and also that only the *green* parts of the plant carried out this process.

Jean Senebier, a Swiss minister, confirmed the findings of Ingenhousz and observed further that plants used as nourishment *carbon dioxide* 'dissolved in water'. Early in the nineteenth century another Swiss scholar, de Saussure, studied the quantitative relationships between the CO_2 taken up by a plant and the amount of organic matter and O_2 produced and came to the conclusion that *water* was also consumed by plants during assimilation of CO_2. In 1817 two French chemists, Pelletier and Caventou, isolated the green substance in leaves and named it *chlorophyll*. Another milestone in the history of photosynthesis was the enunciation in 1845 by Robert Mayer, a German physician, that plants transform energy of sunlight into chemical *energy*. By the middle of the last century the phenomenon of photosynthesis could be represented by the relationship:

$$CO_2 + H_2O + light \xrightarrow[\text{plant}]{\text{green}} O_2 + \text{organic matter} + \text{chemical energy}$$

Accurate determinations of the ratio of CO_2 consumed to O_2 evolved during photosynthesis were carried out by the French plant physiologist Boussingault. He found in 1864 that the photosynthetic ratio – the volume of O_2 evolved to the volume of CO_2 used up – is almost unity. In the same year the German botanist Sachs (who also discovered plant respiration) demonstrated the formation of *starch* grains during photosynthesis. Sachs kept some green leaves in the dark for some hours to deplete them of their starch content. He then exposed one half of a starch-depleted leaf to light and left the other half in the dark. After some time the whole leaf was exposed to iodine vapour. The illuminated portion of the leaf turned dark violet due to the formation of starch-iodine complex; the other half did not show any colour change.

The direct connection between oxygen evolution and chloroplasts of green leaves and also the correspondence between the action spectrum of photosynthesis and the absorption spectrum of chlorophyll (see Chapter 4) were demonstrated by Engelmann in 1880. He placed a filament of the green alga *Spirogyra*, with its spirally arranged chloroplasts, on a microscope slide together with a suspension of oxygen-requiring, motile bacteria (Fig. 2.1). The slide was kept in a closed chamber in the absence of air and illuminated. Motile bacteria would move towards regions of greater O_2 concentration. After a period of illumination the slide was examined under a microscope and the bacterial population counted. Engelmann found that the bacteria were concentrated around the green bands of the algal filament. In another series of experiments he illuminated the alga with a spectrum of light by interposing a prism between the light source and the microscope stage. The largest number of bacteria surrounded those parts of the algal filament that were in the blue and red regions of the spectrum. The chlorophylls present in the alga absorbed blue and red light; since light has to be absorbed to bring about photosynthesis Engelmann concluded that chlorophylls are the active photoreceptive *pigments* for photosynthesis. The state of knowledge on photosynthesis at the beginning of this century could be represented by the equation:

$$(CO_2)_n + H_2O + light \xrightarrow[\text{plant}]{\text{green}} (O_2)_n + \text{starch} + \text{chemical energy}$$

2.2   Further work related to techniques

Though by the beginning of this century the overall reaction of photosynthesis was known, the discipline of biochemistry had not advanced enough to understand the mechanism of reduction of carbon dioxide to carbohydrates. It should be admitted that even now we know very little about certain aspects of photosynthesis. Early attempts were made to study the effects of light intensity, temperature, carbon dioxide concentration, etc. on the overall yields of photosynthesis. Though plants of divergent species were used in these studies, most

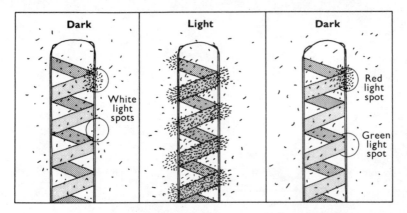

Fig. 2.1 Summary of Engelmann's experiment for studying photosynthesis using the alga *Spirogyra* and motile bacteria. The alga has a spiral chloroplast and the bacteria migrate towards regions of higher O_2 concentration. **Left**: Illumination with a spot of white light. **Centre**: Complete illumination with white light. **Right**: Illumination with spots of red and green light. Note the lack of O_2 evolution in green light.

of the determinations were carried out with unicellular green algae, *Chlorella* and *Scenedesmus*, and the unicellular flagellate *Euglena*. Unicellular plants are more suitable for quantitative research since they can be grown in all laboratories under fairly standard conditions. They can be suspended uniformly in aqueous buffer solutions and aliquots of the suspension can be transferred with a pipette as though they were true solutions. Chloroplasts are best prepared and studied from leaves of higher plants, the most common being spinach leaves as they are usually available fresh in the market and can be grown quite easily; peas and lettuce are also sometimes used.

Since CO_2 is fairly soluble and O_2 is relatively insoluble in water, during photosynthesis in a closed system there will be a change in gas pressure. The Warburg respirometer (adapted by Otto Warburg in 1920) is used in studies involving the action of light on photosynthetic systems by measuring the changes in the O_2 volume of the system (see *Manometric Techniques*, Umbrett, Burris and Stauffer, Burgess Publishing Company, USA, for details).

The oxygen electrode (Fig. 2.2) is a more convenient instrument to measure uptake or liberation of O_2 during a reaction. The electrode works on the principle of polarography and is sensitive enough to detect O_2 concentrations of the order of 10^{-8} moles cm^{-3} (0.01 millimolar). The apparatus consists of platinum wire sealed in plastic as cathode, and an anode of circular silver wire bathed in a saturated KCl solution. The electrodes are separated from the reaction mixture by an O_2 gas-permeable teflon membrane. The reaction mixture in the plastic (or glass) container is stirred constantly with a small magnetic stirring rod. When a voltage is applied across the two electrodes, with the platinum electrode negative to the reference electrode, the oxygen in the solution undergoes electrolytic reduction. The flow of current in the system between 0.5 and

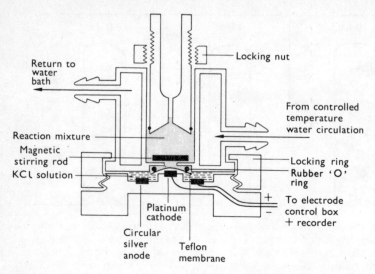

Return to
water
bath

Locking nut

From controlled
temperature
water circulation

Reaction mixture

Magnetic
stirring rod

KCl solution

Locking ring

Rubber 'O'
ring

+
–

To electrode
control box
+ recorder

Platinum
cathode

Circular
silver
anode

Teflon
membrane

Fig. 2.2   The oxygen electrode (Rank Bros., Bottisham, Cambridge).

0.8 V varies in a linear relationship to the partial pressure of the oxygen in solution. The instrument is usually operated at a voltage of about 0.6 V. The current liberated is measured by connecting the electrode to a suitable recorder. The whole apparatus is kept at a constant temperature by circulating water from a controlled temperature water source. The effects of light and of various chemicals on photosynthesis are measured using the oxygen electrode. The O_2 electrode has the advantage over the Warburg method in that rapid and continuous measurements of O_2 evolution can be made. However, the Warburg apparatus can measure up to 20 reaction mixtures simultaneously while the O_2 electrode measures reactions one at a time.

2.3   Limiting factors

The extent of photosynthesis performed by a plant depends on a number of internal and external factors. The chief internal factors are the structure of the leaf and its chlorophyll content, the accumulation within the chloroplasts of the products of photosynthesis, the influence of enzymes and the presence of minute amounts of mineral constituents. The external factors are the quality and quantity of light incident on the leaves, the ambient temperature and the concentration of carbon dioxide and oxygen in the surrounding atmosphere and more generally the availability of water and nutrients.

The effect of light intensity

The effect of light intensity on the photosynthetic activity of a healthy suspension of *Chlorella* cells is illustrated in Fig. 2.3. At low light intensities

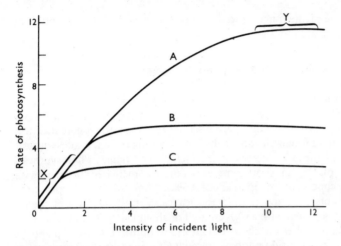

Fig. 2.3 Effect of external factors on rate of photosynthesis. **A**, Effect of light intensity at 25°C and 0.4% CO_2; **B**, at 15°C and 0.4% CO_2; **C**, at 25°C and 0.01% CO_2. All units in the graph are arbitrary.

the rate of photosynthesis, as measured by oxygen evolution, increases linearly in proportion to light intensity. This region of the curve, marked X, is known as the light-limiting region. With more and more light intensity, photosynthesis under atmospheric conditions (21% O_2 and 0.034% CO_2) becomes less efficient until after about 10 000 lux (about 100 W m^{-2}) increasing light intensity produces no further effect on the rate of photosynthesis. This is indicated by the horizontal parts of the curves in the figure. This plateau region designated Y is the light saturation region. If the rate of photosynthesis is to be raised in this region, factors other than light intensity would have to be adjusted. The amount of sunlight falling on a clear summer day in many places on the earth is about 100 000 lux, also equal to about 1000 Wm^{-2}. Thus, except for plants growing in thick forests and in shade, there is often sufficient sunlight incident on the plants to saturate their photosynthetic capacity. The energy of the extreme blue (400 nm) and red (700 nm) light quanta differs only by a factor of two and the photons in this wavelength range are qualitatively efficient to start photosynthesis though, as we shall later see, the leaf pigments preferentially absorb light of certain definite wavelengths.

Effect of temperature

A comparison of the curves A and B in the figure shows that at low light intensities the rate of photosynthesis is the same at 15°C and at 25°C. The reactions in the light-limiting region, like true photochemical reactions, are not sensitive to temperature. At higher light intensities, however, the rate of photosynthesis is much higher at 25°C than at 15°C. Thus, factors other than mere photon absorption influence photosynthesis in the light-saturation region. Most temperate climate plants function well between 10°C and 35°C, the optimum temperature being around 25°C.

Effect of CO_2 concentration: CO_2 compensation point

In the light-limiting region the rate of photosynthesis is not affected by lowering the CO_2 concentration, as shown by curve C in Fig. 2.3. Thus, it can be inferred that CO_2 does not participate directly in the photochemical reaction. But at light intensities above the light-limiting region, photosynthesis is appreciably enhanced by increasing the CO_2 concentration. Photosynthesis by some crop plants, in short-term experiments, increased linearly with increasing CO_2 concentration up to about 0.5% though continued exposure to this high CO_2 concentration injured the leaves. Very good rates of photosynthesis can be obtained with a CO_2 content of about 0.1%. The average CO_2 content of the atmosphere is about 0.035% (350 ppm). Therefore plants in their normal environment do not have enough CO_2 to make maximum use of sunlight falling on them. When a leaf kept in an enclosed space is illuminated with saturating light the concentration of CO_2 in the air falls gradually and reaches a constant level known as the CO_2 compensation point. At this point the CO_2 uptake by photosynthesis is balanced by the CO_2 released by respiration (dark and light). The CO_2 compensation point varies with species; it is very low (less than 10 ppm) for the C_4 plants (Chapter 6) whereas for the C_3 photosynthesis species the value is greater than 50 ppm.

2.4   Light and dark reactions; flashing light experiments

As early as 1905 the British plant physiologist F.F. Blackman interpreted the shape of the light saturation curves by suggesting that photosynthesis is a two-step mechanism involving a photochemical or light reaction and a non-photochemical or dark reaction. The dark reaction which is enzymatic is slower than the light reaction and hence at high light intensities the rate of photosynthesis is entirely dependent upon the rate of the dark reaction. The light reaction has a low or zero temperature coefficient while the dark reaction has a high temperature coefficient, characteristic of enzymatic reactions. It should be clearly understood that the so-called dark reaction can proceed both in light and in darkness.

The light and dark reactions can be separated by using flash illuminations lasting fractions of a second. Light flashes lasting less than a millisecond (10^{-3} s) can be produced either mechanically by placing a slit in a rotating disc in the path of a steady light beam, or electrically by loading up a condenser and discharging it through a vacuum tube. Ruby lasers emitting red light at 694 nm are also used as a radiation source. In 1932 Emerson and Arnold illuminated suspensions of *Chlorella* cells with condenser flashes lasting about 10^{-5} s. They measured the rate of oxygen evolution in relation to the energy of the flashes, the duration of the dark intervals between the flashes, and the temperature of the cell suspension. Flash saturation occurred in normal cells when one molecule of O_2 evolved from 2500 chlorophyll molecules. Emerson and Arnold concluded that the maximum yield of photosynthesis is not determined by the number of chlorophyll molecules capturing the light but by the number of enzyme molecules which carry out the dark reaction. They also observed that for dark time intervals (between successive flashes) greater than 0.06 s the yield of oxygen per flash was independent of the dark time interval; the yield per light flash increased with dark time intervals from 0 to 0.06 s. Thus the dark reaction determining the saturation rate of photosynthesis takes about 0.06 s for completion. The average dark reaction time was calculated to be about 0.02 s at 25°C.

2.5 Important discoveries and formulations

The state of knowledge in the field of photosynthesis at the turn of this century could have been represented by the equation:

$$CO_2 + H_2O \xrightarrow[\text{chlorophyll}]{\text{light}} (CH_2O) + O_2 \ (\Delta G = 48 \times 10^4 \text{ J (114 kcal)})$$

Prior to about 1930 many investigators in the field believed that the primary reaction in photosynthesis was splitting of carbon dioxide by light to carbon and oxygen; the carbon was subsequently reduced to carbohydrates by water in a different series of reactions. Two important discoveries in the 1930s changed this viewpoint. Firstly, a variety of bacterial cells were found to assimilate CO_2 and synthesize carbohydrates without the use of light energy. Then the Dutch microbiologist van Niel in comparative studies of plant and bacterial photosynthesis showed that some bacteria can assimilate CO_2 in light without evolving O_2. Such bacteria would not grow photosynthetically unless they were supplied with a suitable hydrogen donor substrate. Photosynthesis could be represented, according to van Niel, by the general equation:

$$CO_2 + 2H_2A \xrightarrow[\text{chlorophyll}]{\text{light}} (CH_2O) + H_2O + 2A$$

where H_2A is the oxidizable substrate. Van Niel suggested that photosynthesis of green plants and algae is a special case in which H_2A is H_2O and 2A is O_2. The primary photochemical act in plant photosynthesis would be the splitting of water to yield an oxidant (OH) and a reductant (H). The primary reductant (H) could then bring about the reduction of CO_2 to cell materials and the primary oxidant (OH) could be eliminated through a reaction to liberate O_2 and reform H_2O. The overall equation of photosynthesis for green plants, after van Niel, is:

$$CO_2 + 4H_2O \xrightarrow[\text{chlorophyll}]{\text{light}} (CH_2O) + 3H_2O + O_2$$

which is a sum of three individual steps:

$$(i)\ 4H_2O \xrightarrow[\text{green pigments}]{\text{light}} 4(OH) + 4H$$

$$(ii)\ 4H + CO_2 \rightarrow (CH_2O) + H_2O$$

$$(iii)\ 4(OH) \rightarrow 2H_2O + O_2$$

The reaction sequences clearly show that the oxygen is evolved from water and not from CO_2.

The second important observation was made in 1937 by R. Hill of Cambridge University. Hill separated the photosynthesizing particles (chloroplasts) of green leaves from the respiratory particles by differential centrifugation of a homogenate of leaf tissues. Hill's chloroplasts did not evolve O_2 when illuminated as such (due to possible damage of the chloroplasts during isolation) but did so when suitable electron acceptors (oxidants) like potassium ferrioxalate or potassium ferricyanide were added to the illuminated suspension. One molecule of O_2 was evolved for every four equivalents of oxidant reduced photochemically. Later many quinones and dyes were found to be reduced by illuminated chloroplasts. The chloroplasts, however, failed to reduce CO_2, the natural electron acceptor of photosynthesis. This phenomenon, now known as the Hill reaction, is a light-driven transfer of electrons from water to non-physiological oxidants (Hill reagents) against the chemical potential gradient. The significance of the Hill reaction lies in the demonstration of the fact that photochemical O_2 evolution can be separated from CO_2 reduction in photosynthesis.

The decomposition of water, and the resulting liberation of O_2 during photosynthesis, was established by Ruben and Kamen in California in 1941. They exposed photosynthesizing cells to water enriched in the recently available oxygen isotope of mass 18 (^{18}O). The isotopic composition of the oxygen evolved was the same as that of the water and not that of CO_2 used. Kamen and Ruben also discovered the radioactive isotope ^{14}C which was successfully used by Bassham, Benson and Calvin in California to trace the path of carbon in photosynthesis (Chapter 6). Calvin and co-workers showed that reduction of CO_2 to sugars proceeded by dark enzymic reactions and also that two molecules of reduced pyridine nucleotide [NADPH$_2$] and three molecules of ATP were required for the reduction of every molecule of CO_2. The role of pyridine nucleotides and ATP in the respiration of tissues was already established by this time. The photosynthetic reduction of NADP to NADPH$_2$ by isolated chloroplasts with simultaneous evolution of O_2 was demonstrated in 1951 by three different laboratories. Arnon, Allen and Whatley in 1954 also demonstrated cell-free photosynthesis, i.e. the assimilation of CO_2 and evolution of O_2 by isolated spinach chloroplasts. Proteins like ferredoxin, plastocyanin, ferredoxin-NADP reductase and cytochromes b and f, which participate in the transfer of electrons in photosynthesis, were isolated from chloroplasts within a decade.

Emerson and coworkers measured quantum yields of photosynthesis in algae and observed that the average quantum yield obtained by using two super-

using two superimposed beams of different wavelengths was higher than the average quantum yield obtained by using the two light beams separately. To explain this enhancement of quantum yield Emerson and Rabinowitch in 1960 postulated the existence of two light reactions in photosynthesis. In the same year Hill and Bendall put forward the Z scheme of photosynthesis showing the operation of two photosystems in series in photosynthetic electron transport and phosphorylation.

To conclude, healthy green leaves on illumination generate $NADPH_2$ and ATP. The reducing power of $NADPH_2$ and the energy of hydrolysis of ATP are utilized to reduce CO_2 to carbohydrates in the presence of enzymes some of whose activities are regulated by light.

3

Photosynthetic Apparatus

The photosynthetic apparatus is that part of the leaf or algal cell which contains the ingredients for absorbing light and for channelling the energy of the excited pigment molecules into a series of chemical and enzymatic reactions. Engelmann's experiments (Chapter 2) have shown that chlorophylls are the pigments responsible for capturing light quanta. Knowledge concerning the sub-cellular structure in which chlorophyll is located comes from light and electron microscopy, and from cell fractionation techniques. In green algae and in higher plants the chlorophyll is contained in a cellular plastid called the *chloroplast*.

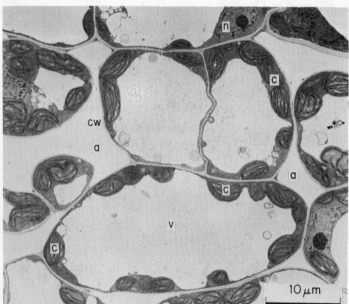

Fig. 3.1 Thin section of spinach mesophyll cells showing chloroplasts (**c**) in cytoplasm extending around the inside of the cell wall (**cw**). **n**, nucleus; **v**, vacuole; **a**, air space between cells allowing easy diffusion of gases to chloroplasts. (Courtesy A. D. Greenwood, Imperial College, London.)

Electron microscope pictures show that chloroplasts in higher plants, e.g. spinach, tobacco, are saucer-shaped bodies 4 to 10 μm in diameter and 1 μm in thickness (1 μm = 10^{-6} m) with an outer membrane or envelope separating it from the rest of the cytoplasm (Fig. 3.1). The number of chloroplasts per cell, in higher plants, varies from one to more than a hundred depending upon the particular plant and on the growth conditions. In many plants the chloroplasts are able to reproduce themselves by a simple division.

Internally the chloroplast is comprised of a system of lamellae or flattened *thylakoids* which are arranged in stacks in dense green regions known as *grana* (Fig. 3.2). Each lamella in the chloroplast may contain two double-layer membranes. The grana are embedded in a colourless matrix called the *stroma* and the whole chloroplast is surrounded by a bounding double membrane, the *chloroplast envelope*. Within the chloroplast the grana are interconnected by a system of loosely arranged membranes called the stroma lamellae. The detailed structure of the thylakoids is shown in Fig. 3.3. These models are based on electron microscopy using freeze-fracturing techniques. The surfaces thus exposed show the distribution of chlorophyll-protein complexes embedded in or associated with the lipid bilayer which forms the 'backbone' of the membrane. The organization of these complexes which are seen as particles is different in the stacked, grana membrane regions than in the unstacked, stroma membranes. This is explained by predominance of the photosystem II, oxygen evolving reaction complexes in the grana and the photosystem I particles in the stroma lamellae (see Fig. 8.3, p. 101).

The lamellar structure is found not only in chloroplasts of higher plants but also in algal chloroplasts. In algae the shapes of the chloroplast are varied. Figure 3.4 shows chloroplasts from a green and a red alga and the cell structure of a blue-green alga. The most primitive algae, the blue-greens (also called Cyanobacteria since they are prokaryotic organisms) do not contain chloroplasts as such. The photosynthetic material present in these organisms consists of parallel layers of lamellar membranes traversing the cytoplasm.

It is possible to fractionate the chloroplasts of higher plants so that the green lamellae are separated from the colourless stroma matrix. The lamellar membranes in which the chlorophyll is embedded are approximately half lipid and half protein in chemical composition. The proteins catalyze the enzyme reactions and give mechanical strength to the membranes. Most of the light-harvesting chlorophylls *a* and *b* are conjugated with specific membrane proteins. The presence of lipids in the membranes facilitates energy storage and offers selective permeability of sugars, salts, substrates, etc. Chloroplast lipids play an important role in maintaining membrane structure and function. One of the causes for the thermal and photo decay of chloroplasts is the release of lipids from the membranes and their oxidation.

The quantum conversion of light energy and associated electron transport reactions of photosynthesis occur in the lamellae. The stroma contains many soluble proteins including the enzymes of the Calvin cycle (Chapter 6) which carry out the dark phase reduction of CO_2 to carbohydrates.

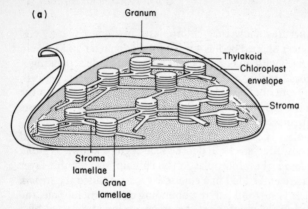

(a)

Granum

Thylakoid

Chloroplast
envelope

Stroma

Stroma
lamellae

Grana
lamellae

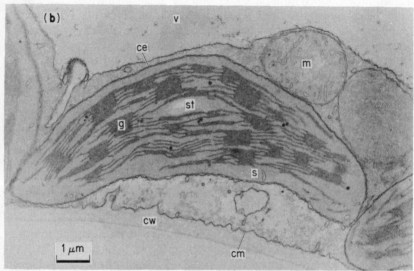

(b)

v

ce

m

st

g

s

cw

cm

1 μm

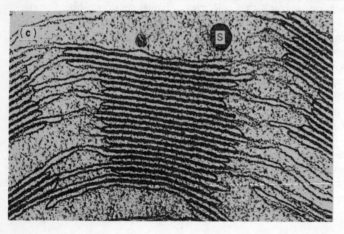

(c)

S

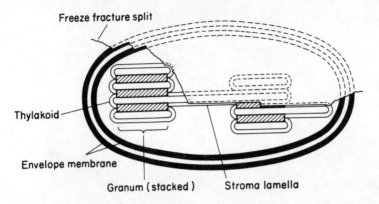

Fig. 3.3 Diagram of chloroplast structure as revealed by freeze etching in the electron microscope. Freeze fracturing of the chloroplast results in cross-fracturing or splitting of the lamellae, thus exposing interior regions of the membrane (after Staehelin, L. A. (1976). *Journal of Cell Biology*, **71**, 136). A more detailed recent description is shown in Figs 8.1 and 8.3.

3.1   Isolation of chloroplasts from leaves

The first photosynthetically active chloroplasts were isolated by Hill; these preparations were active only in O_2 evolution coupled to the reduction of non-physiological electron acceptors (see Chapter 2). Arnon and Whatley isolated chloroplasts in isotonic sodium chloride, i.e. about 0.35 M or 2%. Their preparations were capable of photoreduction of NADP and photophosphorylation but were able to fix CO_2 only at low rates, although they contained all the enzymes of the Calvin CO_2 fixation cycle. These chloroplasts appeared intact under the light microscope but electron microscope pictures of the preparation indicated that they had lost their outer membranes and were naked lamellar systems. Such preparations are called 'Type C' (broken) chloroplasts (Fig. 3.5). Walker has developed techniques for the isolation of 'Type A' (complete) chloroplasts which retain their outer envelopes and which can fix CO_2 at rates up to 90% of those of whole leaves. See Hall (1972) *Nature*, **235**, 125 for discussion of chloroplast Types.

Two methods used in our laboratory for the preparation of chloroplasts are given below. All solutions and apparatus used should be pre-cooled in ice. The preparation should be carried out as quickly as possible. See also Chapter 5.

Fig. 3.2   *(left)* **(a)** Cut-away representation of a chloroplast to show three-dimensional structure. **(b)** Section of a chloroplast in the cytoplasm of a spinach leaf cell. **ce**, chloroplast envelope; **g**, granum consisting of stacks of thylakoids; **s**, stroma; **st**, starch granule in chloroplast; **cm**, cytoplasmic membrane; **cw**, cell wall; **m**, mitochondrion; **v**, vacuole. **(c)** A single granum within a chloroplast showing stacks of thylakoids and interconnecting stroma lamellae between granal stacks; **s**, lipid droplet in the stroma. (Courtesy of A. D. Greenwood.)

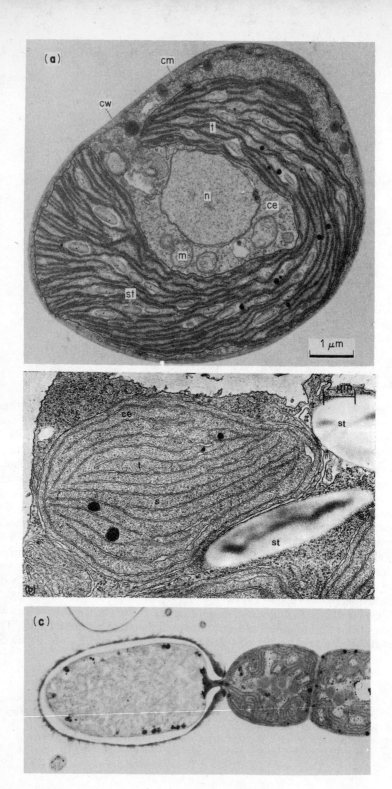

(a) cm cw ti n ce m st 1 μm

(b) ce ti s st st μm

(c)

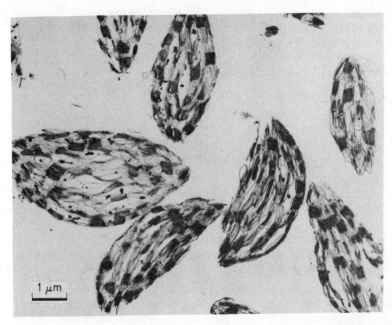

Fig. 3.5 Bean chloroplasts isolated in buffered sucrose media showing chloroplasts with intact thylakoids but without a chloroplast envelope – Type C as defined below. (Courtesy of A. D. Greenwood and R. Leech.)

Preparation 1 Procedure of Whatley and Arnon (modified).
Grinding medium: 0.35 M NaCl; 0.04 M tris/HCl buffer pH 8.0

Cut 25 g of spinach (*Spinacia oleraceae*) leaves into small pieces 0.5–1 cm long. Place in a laboratory or domestic blender with 50 cm³ grinding medium. Blend for 10 s on low and 20 s on high speed. Filter the homogenate through a nylon bag (or 4 layers of cheesecloth) into a centrifuge tube. Centrifuge 4 min at 2000 g. Discard the supernatant. Re-suspend the pellet in 2 cm³ 0.35 M NaCl using a small piece of absorbent cotton wool wrapped round the end of a glass

Fig. 3.4 (a) The green alga *Coccomyxa* sp., which is a symbiont within a lichen, showing a single cup-shaped chloroplast within the cell. The thylakoids (**t**) are in groups of three. **ce**, chloroplast envelope; **st**, starch granule within chloroplast; **cw**, cell wall; **cm**, cytoplasmic membrane; **n**, nucleus; **m**, mitochondrion. (Courtesy of H. Bronwen Griffiths, Imperial College, London.) (b) Chloroplast of the red alga *Ceramium* sp. with single thylakoids (**t**) lying nearly parallel to one another in the stroma (**s**). **ce**, chloroplast envelope; **st**, starch granule outside chloroplast. Dark spots in chloroplast are lipid droplets. (Courtesy of A. D. Greenwood.) (c) Vegetative and heterocyst cells of the cyanobacterium *Anabaena azollae*, a symbiont in the *Azolla* fern from rice fields. X 5000 (Courtesy of D. J. Shi, King's College, London.)

rod. This preparation consists of Type C (broken) chloroplasts. Chloroplast fragments (Type E) are prepared by diluting the suspension with 10 vol of water to give a NaCl concentration of 0.035 M.

Preparation 2   Procedure of Walker.
Grinding medium:

Sorbitol	0.33 M
$MgCl_2$	0.005 M
$Na_4P_2O_7.10H_2O$	0.01 M

Adjust pH of above mixture to 6.5 with HCl and add sodium isoascorbate to a final concentration of 0.002 M just before use.

Re-suspending medium:

Sorbitol	0.33 M
$MgCl_2$	0.001 M
$MnCl_2$	0.001 M

EDTA (ethylenediamine tetra acetate) 0.002 M.
HEPES (hydroxyethylpiperazine-ethanesulphonic acid – buffer) 0.05 M.
Adjust pH to 7.6 with NaOH.

Homogenize 50 g of chilled spinach leaves with 200 cm^3 of freshly-made grinding medium for 3 to 5 s in a domestic blender. Squeeze the macerate through 2 layers of cheesecloth and filter through 8 layers of cheesecloth into 50 cm^3 plastic centrifuge tubes. Centrifuge rapidly at 0°C from rest to 4000 g to rest in approximately 90 s. Re-suspend the pellet gently using a glass rod and a small piece of absorbent cotton in 1 cm^3 of resuspending medium. This procedure should produce a suspension of chloroplasts, 50–80%, Type A (complete), which would be capable of high rates of CO_2 fixation.

Notes   The yield of chloroplasts is increased if the leaves are floated on cold running water or in an ice bath and brightly illuminated for about 30 minutes prior to grinding. To avoid too much starch in the leaves spinach should be harvested early in the morning – the presence of white rings in the chloroplast pellet is indicative of excessive starch and calcium oxalate in the spinach and results in the disruption of the chloroplasts. A number of species other than *Spinacia oleracea* are often called spinach (e.g. beet and Swiss chard); however these do not yield chloroplasts with good CO_2 fixation rates. Peas, lettuce and *Chenopodium album* are acceptable substitutes for spinach.

3.2   Chloroplast pigments

All photosynthetic organisms contain one or more organic pigments capable of absorbing visible radiation which will initiate the photochemical reactions of photosynthesis. These pigments can be extracted from most leaves into alcohol or into other organic solvents. From the alcoholic extract individual pigments can be separated by chromatography on a column of powdered sugar, as was shown by the Russian botanist Tswett in 1906. The three major classes of pigments found in plants and algae are the chlorophylls, the carotenoids and the phycobilins. The chlorophylls and carotenoids are insoluble in water but the phycobilins are soluble in water. The carotenoids and phycobilins are called the accessory photosynthetic pigments since the quanta absorbed by these pigments can be transferred to chlorophyll. Table 3.1 gives the absorption characteristics of these pigments. The photosynthetic pigments of bacteria are discussed in Chapter 7.

Chlorophylls are the pigments that give plants their characteristic green colour. They are insoluble in water but soluble in organic solvents. Chlorophyll *a* is bluish-green and chlorophyll *b* is yellowish-green. Chlorophyll *a* is present in all photosynthetic organisms which evolve O_2. Chlorophyll *b* is present (about one third of the content of chlorophyll *a*) in leaves of higher plants and in green algae. The absorption maxima of chlorophyll *a* and chlorophyll *b* in ether are respectively at 660 and 643 nm as shown in Fig. 3.6; in acetone the peaks are at 663 and 645 nm. However, careful spectroscopic investigations of the living cell indicate the presence of multiple forms of chlorophyll *a* in vivo. These forms of chlorophyll *a* may be associated in different ways with the lamellae and have different photochemical functions.

The molecular formula for chlorophyll *a* is $C_{55}H_{72}N_4O_5Mg$ and for chlorophyll *b* is $C_{55}H_{70}N_4O_6Mg$. The structural formula of chlorophyll was determined by Fischer in Germany in 1940 from degradative studies; the structure was confirmed by the complete synthesis of the molecule by Woodward at Harvard in 1960. The chlorophyll molecule (Fig. 3.7) contains a porphyrin 'head' and a phytol 'tail'. The polar (water soluble) porphyrin nucleus is made

Table 3.1   The pigments.

Type of pigment	Characteristic absorption maxima (nm) (in organic solvents)	Occurrence
Chlorophylls		
Chlorophyll *a*	420, 660	All higher plants and algae
Chlorophyll *b*	435, 643	All higher plants and green algae
Chlorophyll *c*	445, 625	Diatoms and brown algae
Chlorophyll *d*	450, 690	Red algae
Carotenoids		
β-carotene	425, 450, 480	Higher plants and most algae
α-carotene	420, 440, 470	Most plants and some algae
Luteol	425, 445, 475	Green algae, red algae and higher plants
Violaxanthol	425, 450, 475	Higher plants
Fucoxanthol	425, 450, 475	Diatoms and brown algae
Phycobilins		
Phycoerythrins	490, 546, 576	Red algae and in some blue-green algae (cyanobacteria)
Phycocyanins	618	Blue-green algae and in some red algae
Allophycocyanins	650	Blue-green and red algae

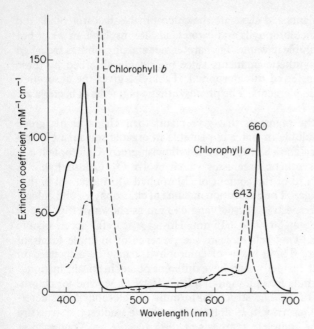

Fig. 3.6 Absorption spectra of chlorophylls extracted in ether.

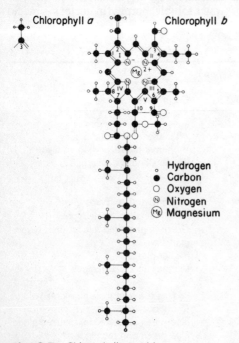

Fig. 3.7 Chlorophyll *a* and *b* structure.

up of a tetrapyrrole ring and a magnesium atom. In the cell, electron micro-scopists think that the chlorophyll is sandwiched between protein and lipid layers of the chloroplast lamellae. The porphyrin part of the molecule is bound to the protein while the phytol chain extends into the lipid layer since it is soluble in lipids. Pheophytins, which are chlorophylls without the central Mg atom, have been recently identified as constituents of the photosynthetic electron transport chain.

The optical absorption curves of chlorophyll *a* and chlorophyll *b* intersect at 652 nm. A solution of chlorophyll at a concentration of 1 mg cm^{-3} has an absorbance of 34.5 at 652 nm. Arnon has devised the following method for the determination of the chlorophyll content of a chloroplast suspension by measuring its absorption at 652 nm. Dilute 0.1 cm^3 of chloroplast suspension with 20 cm^3 of 80% acetone, mix, and filter. Read the absorbance of the filtrate at 652 nm in a spectrophotometer in a 1 cm light path cell against 80% acetone as reference. Multiply the absorbance by 5.8 to give mg of chlorophyll per cm^3 of the original chloroplast suspension.

The carotenoids are yellow or orange pigments found in all photo-synthesizing cells. Their colour in the leaves is normally masked by chloro-phyll, but in the autumn season when chlorophyll disintegrates the yellow pigments become visible. Carotenoids contain a conjugated double bond system of the polyene type. They are usually either hydrocarbons (carotenes) or oxygenated hydrocarbons (carotenols or xanthophylls) of 40 carbon chains built up from isoprene subunits (Fig. 3.8). They have triple-banded absorption spectra in the region from about 400 to 550 nm. The carotenoids are situated in the chloroplast lamellae in close proximity to the chlorophyll. The energy absorbed by the carotenoids may be transferred to chlorophyll *a* for photo-synthesis. In addition the carotenoids may protect the chlorophyll molecules from too much photo-oxidation in excessive light.

Blue-green algae and red marine algae contain a group of pigments known as *phycobilins* (Fig. 3.9). Phycobilins are linear tetrapyrroles structurally related to chlorophyll *a* but they do not have the phytyl side chain, nor do they contain magnesium. The chromophores of phycobilins are covalently linked to poly-peptides to form water-soluble phycobiliproteins. There are three classes of phycobilins viz. *phycoerythrins*, *phycocyanins* and *allophycocyanins*. The red phycoerythrins found in all red algae (Table 3.1) absorb light in the middle of the visible spectrum. This enables the red algae living under the sea to perform photosynthesis in the dim bluish-green light reaching the lower surfaces of the ocean – the deeper under the sea a red alga lives the more phycoerythrin it contains in relation to chlorophyll. The blue phycocyanins and allophyco-cyanins occur in the blue-green algae which live on the surface layers of lakes and on land. The energy absorbed by the phycobilins (accessory pigments) is transferred to the chlorophyll for photochemical processes. Using pico (10^{-12}) second spectroscopy it has been shown that the energy transfer in the red alga, *Porphyridium cruentum* occurs in the sequence:

phycoerythrin $\rightarrow$ phycocyanin $\rightarrow$ allophycocyanin $\rightarrow$ chlorophyll *a*.

Thus higher plants and algae, during the course of evolution, have developed

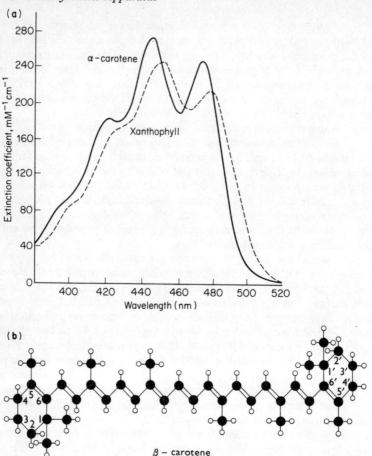

Fig. 3.8   (a) Absorption spectra of α-carotene and a xanthophyll. (b) Structure of β-carotene.

various pigments to capture the available solar radiation most efficiently and to carry out photosynthesis. The relative abundance of these pigments depends upon the species, the location of the plant, the seasons, etc.

In addition to the conjugated pigment-protein complexes which are involved in light harvesting and energy transduction, the chloroplast lamellae also contain many other proteins, lipids, quinones, and ions. The oxygen-evolving complex, for example, is thought to be a 34 kilodalton (34 kD) manganese protein which is attached to three other polypeptides of molecular weight 18, 23 and 33 kilodaltons (Fig. 5.2b, p. 55). The secondary quinone acceptor Q_B of photosystem II is conjugated to a 32 kD polypeptide which can also bind certain herbicides and bicarbonate. Three cytochromes, b_{559}, b_6 and f, have been identified as chloroplast components by differential optical absorption spectro-

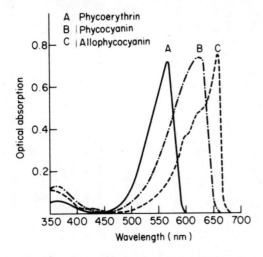

Fig. 3.9 Absorption spectra of phycobiliproteins in the visible region.

scopy. The blue copper protein plastocyanin, the iron-sulphur protein ferredoxin and the flavoprotein, ferredoxin-NADP reductase, all play important
roles in photosynthesis. The properties of these and other chloroplast electron
transport chain components are summarized in Table 3.2 and will be discussed
in more detail in later chapters.

Plastoquinones (quinones in plastids or chloroplasts) are the most abundant
electron mediators found in chloroplasts. Plastoquinones possess a dimethyl-
benzoquinone ring to which long chain aliphatic hydrocarbons (phytyl) are
attached. The structure of plastoquinone A is:

$$H_3C \overset{\displaystyle C}{\underset{\displaystyle O}{\bigcirc}} (CH_2 - CH = \overset{\displaystyle CH_3}{\underset{\displaystyle}{C}} - CH_2)_9 - H$$

The unsubstituted quinone, benzoquinone, and dimethyl benzoquinone are
used as artificial electron acceptors (Hill reagents) of photosystem II. The
quinone ring can be reduced to a semiquinone free radical (addition of an
electron and a proton) or to a quinol (addition of two protons and electrons).

The *lipids* play a dynamic role in the function of the thylakoid membrane
components. The more abundant lipids in the bilayer matrix of the membrane
are the glycolipids, monogalactosyl diacylglycerol and digalactosyl
diacylglycerol, which constitute about 75% of the membrane lipids. Recent
studies using tritium labelling (exchange of H atoms with H^3) of right side-
out and inside-out membrane vesicles (Chapter 4) indicate an asymmetric

Table 3.2 Properties of chloroplast electron transfer chain components (see also Fig. 5.2 pp 54 & 55).

Component and symbol	Molecular mass M_r: daltons	E_m: mid point redox potential volts (see §5.1)	Detection, probable function, etc.
'Water-oxidizing' protein M	34 000	+ 0.82?	Reconstitution studies; inner membrane; associated with 2 or 4 Mn atoms and surrounded by 3 polypeptides of M_r 33, 23 and 18 kD, and Cyt b 559; oxidizes H_2O donating electrons to Z and releasing protons into the lumen of thylakoid.
Primary e^- donor to PSII (Z)		+ 1.0	EPR (electron paramagnetic resonance), site-directed mutagenesis; may be a tyrosine bound to the reaction centre (RC); $1e^-$ mediator between M and P680.
P680	1 or 2 × 892	+ 1.0 − 1.0 (excited state)	Flash absorption spectroscopy and EPR; special Chl a monomer or dimer bound to the $D_1 - D_2$ protein or to 47 kD RC protein; energy trap of PSII.
Pheophytin (pheo)	868	− 0.614	EPR; metastable intermediate e^- acceptor from P*680
Plastoquinones Q_A (or Q_1)	740 (PQ A)	− 150	UV absorption or fluorescence; first stable e^- acceptor; $1e^-$ mediator between pheo and Q_B; may be bound to 32 kD protein D_2.
Q_B (or Q_2)			Flash absorption; two e^- gate between Q_A and PQ pool; reduced to semiquinone and then to quinol; semiquinone bound to 32 kD protein D_1 which can also bind herbicides or HCO_3^-.
Cytochrome b_{559}	9 000 & 4000 Two sub-units	+ 0.080 low potential + 0.380 high potential	Absorption spectroscopy; may be involved in O_2 evolution and cyclic e^- transport around PSII.

PSII complex; predominantly in the grana (appressed membranes)

continued

Cyt b-f complex in grana and stroma lamellae			
Cytochrome b_{563} (or b_6)	23 440	−0.050 high potential −0.170 low potential	Absorption spectroscopy; $PQ \rightarrow PQH_2$ redox cyclic e^- transport; energy transduction; spans the membrane.
Cytochrome f	34 000	+∼0.350	Absorption spectroscopy; c-type cytochrome; $1e^-$ donor to PC.
Rieske iron-sulphur protein $[Fe-S]_R$	20 000	+0.290	EPR; $1e^-$ acceptor from PQ pool.
PSI complex; predominantly in the stroma lamellae (non-appressed membranes)			
P700	1 or 2 × 892	+0.480	Optical absorption, EPR; Chl a monomer or dimer or an epimer of Chl a at C_{10} (in Fig. 3.6); bound to 80 kD protein; energy trap for PSI.
A_0	not known	not determined	Flash EPR; metastable primary e^- acceptor from P*700; may be a Chl a monomer.
A_1	not known	not determined	Flash EPR, reconstitution studies; transient $1e^-$ mediator between A_0 and X; may be a phylloquinone (vitamin k_1)
FX	not known	∼ −0.710	EPR, redox titrations; $1e^-$ mediator to F_A and F_B; may be an [Fe-S] centre bound to P700-Chl a protein.
F_A F_B	not known not known	−0.550 −0.590 $\}$	EPR, redox titrations; $1e^-$ mediators between FX and Fd; Two [4Fe-4S] centres bound to protein 9kD.
Ferredoxin – NADP reductase (FNR)	40 000	−0.380	Isolation from outer lamellae; 1 FAD/mole; e^- mediator between Fd and NADP; can act as diaphorase and a transhydrogenase.

Table 3.2 Continued

	Component and symbol	Molecular mass M_r: daltons	E_m; mid point redox potential volts (see §5.1)	Detection, function, etc.
Mobile e^- carriers	Plastoquinones (PQ)	Variable	~ − 0.100	Optical absorption; abundant in chloroplasts; shuttles e^- and H^+ between PSII and cyt b-f complex; energy transduction and cyclic e^- transport from reduced Fd.
	Plastocyanin (PC)	10 500 per Cu	+ 0.360	Isolation from leaves; 4 or 2 Cu per mole; lumen side of thylakoids; 1e^- mediator between cyt f and P700.
	Ferredoxin (Fd)	11 000	− 0.420	Soluble non-haem iron protein with [2Fe-2S] centre; cyclic and non cyclic e^- transport; donates e^- to various chloroplast constituents.

distribution of these lipids – about 60% of both galactolipids located in the outer half and the rest in the inner half of the bilayer. The minor lipid components – the sulpholipids and phospholipids – are mainly distributed in the inner half of the bilayer.

3.3 The photosynthetic unit

Functionally chlorophyll molecules act in groups. A photosynthetic unit is conceived as a group of pigments and other molecules utilizing the transfer of excitation energy as a mechanism by which the reaction centre communicates with an antenna of light-harvesting pigments as shown in Fig. 3.10. According to this concept a single quantum of energy absorbed anywhere in a set of about 250 to 300 chlorophyll molecules migrates to a reaction centre containing a special pair of chlorophyll a molecules and promotes an electron transfer event. The following observations lead to the idea of the existence of a photosynthetic unit.

1. Approximately 8 quanta of light absorbed by chlorophyll are required for the photosynthetic reduction of 1 CO_2 molecule and the evolution of 1 O_2 molecule. If each chlorophyll molecule in a plant can react photochemically a sufficiently intense flash of light should bring about the evolution of 1 O_2 for every 8 chlorophyll molecules present. However, the flashing light experiments of Emerson and Arnold (Chapter 2) on *Chlorella* suspensions showed that the maximum yield per flash was 1 O_2 molecule for about 2500 chlorophyll molecules, that is a quantum of light is absorbed by one in a cluster of about 300 chlorophyll molecules.

2. Gaffron and Wohl calculated that a single chlorophyll molecule in a dimly illuminated plant will absorb a light quantum only once in several minutes. At this rate a single molecule of chlorophyll will require nearly an hour to capture the light quanta needed for the evolution of one molecule of O_2. But when a plant is illuminated the maximum rates of CO_2 uptake and O_2 evolution are

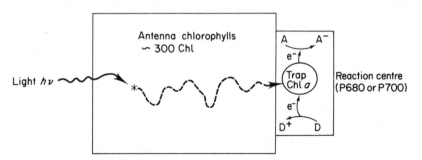

Fig. 3.10 Diagrammatic representation of a photosynthetic unit in chloroplasts. A photon captured by an antenna chlorophyll is transferred to a trap chlorophyll a where photochemical charge separation takes place.

quickly established. So Gaffron and Wohl postulated that the energy harvested by a large set of chlorophyll molecules is conducted to a single reaction centre.

3. Gaffron and co-workers also observed that certain golden-yellow leaves of tobacco, with very little chlorophyll, may reach a nearly normal rate of photosynthesis at very high light intensities. These leaves would have contained a larger proportion of the special type of chlorophyll molecule which is in direct touch with the components of the electron transfer chain.

4. Difference spectroscopy has revealed the role of certain special components like P700 (by Kok in 1956) and cytochrome (by Duysens in 1961), in the photochemical electron transfer reactions. There is one molecule of light-reacting cytochrome and one P700 for every 250 chlorophyll molecules in higher plants and algae.

This concept of a photosynthetic unit is now of historical interest only. It has been revised a number of times in the last twenty-five years, in the light of important discoveries and postulates, and is sure to be modified again to accommodate data emerging from new experimentation. Some of these developments include (*a*) the observation that two photosystems with different reaction centre pigments operate in synchrony in the chloroplasts as proposed in the Z scheme by Hill and Bendall, (*b*) the discovery of the association of oxygen evolution with photosystem II and of ferredoxin reduction with photosystem I, (*c*) the isolation of reaction centres and different membrane complexes, from the photosynthetic membranes and the localization of photosystems and electron transfer chain components in the membranes as a result of electron microscopic and immunological cross reaction studies.

A photosynthetic unit is presently conceived as an integrated assembly of about 600 chlorophyll molecules (300 chlorophylls per reaction centre) and an electron transport chain that could independently harvest light, resulting in oxygen evolution and NADP reduction.

3.4   Photosynthetic apparatus of C_4 plants

Leaves of plants like sugar cane, maize (corn), *Sorghum*, *Amaranthus* and many tropical grasses contain two distinct types of chloroplasts. The leaves of these plants possess 'Kranz-type' anatomy (kranz in German means wreath). The chloroplasts are located in the leaf cells in two concentric layers surrounding the vascular bundle; the inner layer is called the *bundle-sheath* and the outer the *mesophyll*. Chloroplasts from these plants possess a membrane system in the peripheral part of the stroma, called the peripheral reticulum, which connects the thylakoid membranes to the chloroplast envelope. The two types of chloroplasts can be separated by careful grinding and centrifugation using density gradients. As will be discussed in Chapter 6 these plants are able to fix CO_2 by two different pathways: (*a*) the normal Calvin cycle in the bundle-sheath chloroplasts where the initial product of CO_2 fixation is the three carbon compound phosphoglyceric acid (C_3 pathway) and (*b*) in the mesophyll chloroplasts by combining CO_2 with phosphoenol pyruvate to produce the four carbon acids

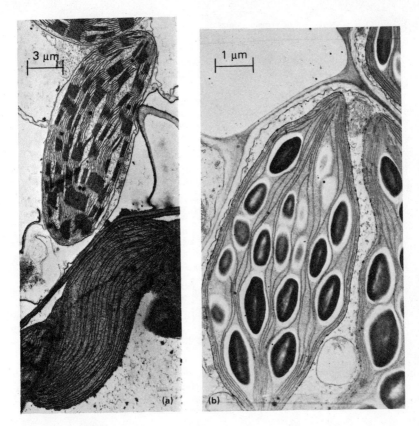

Fig. 3.11 (a) Two different types of chloroplasts in the maize leaf (a C_4 plant). **Above**, mesophyll or granal type. **Below**, bundle sheath or agranal type. There are no starch grains in the bundle sheath chloroplast as the leaf was kept in darkness for 24 hours before fixation. (Courtesy G. Montes, King's College, London.) **(b)** Bundle sheath (agranal) chloroplast from maize. (From Whatley and Whatley, 1980.)

oxaloacetate and malate (C_4 pathway). Plants which fix CO_2 only by the Calvin cycle are termed C_3 species and are generally grown in the temperate zones, e.g. wheat, spinach, oak tree, and those plants which fix CO_2 both by the Calvin cycle and by the malate pathway are termed C_4 species and are generally adapted to growth in warmer and/or drier climates as found in hotter zones.

The mesophyll chloroplasts of C_4 plants (Fig. 3.11a) are randomly distributed in the cell, have stacks of grana and few starch granules. The bundle sheath chloroplasts are relatively larger in size, generally lack grana and possess a number of starch granules (Fig. 3.11b). The mesophyll and bundle sheath chloroplasts lie adjacent in the leaf and photosynthetic products can easily flow

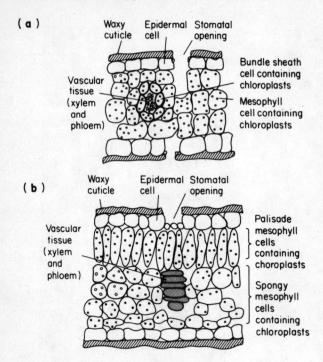

Fig. 3.12 Differences in leaf structure in **(a)** C_4 and **(b)** C_3 species. Note the absence of bundle sheath cells in the C_3 species. (After Zelitch, I., 1979, *Chemical and Engineering News*, *U.S.A.*, Feb. 5, pp. 28–48.)

from one type to the other (see Fig. 3.12a). Generally plants possessing these dimorphic type chloroplasts have a very low CO_2 compensation point, very low photorespiration and glycollate metabolism, and grow more rapidly with higher crop yields.

4

Light Absorption and the Two Photosystems

The most stable states of atoms are those in which the valence electrons are distributed, in accordance with the Pauli principle, into the quantum states of least energy, i.e. the electrons are in their ground states of energy level. When light is absorbed by an atom in the ground state the whole energy of the quantum ($h\nu$) is added to it, and the electrons are lifted to an energy-rich excited state. This is illustrated in Fig. 4.1, taking the helium atom as an example. The time needed for the entire process is of the order of 10^{-15} s. If an atom has an even number of electrons, the spins of these electrons are usually arranged in opposite directions and cancel each other out so that the total electronic spin of the atom s = 0 (singlet state). For an atom with an odd number of electrons the net spin is s = $\frac{1}{2}$ (doublet state). If an atom has an even number of electrons and if the electronic spins are in the parallel direction the net spin is s = 1 and the atom is said to be in the triplet state. These states are illustrated in Fig. 4.1. The transition from the ground to the excited state of an atom by the absorption of a single quantum of energy can be followed by a sharp line in its absorption spectrum at the wavelength λ given by $\Delta E = hc/\lambda$ (see Chapter 1). In a molecule consisting of various atoms the transition from the ground to the excited state can take place by the absorption of light of varying amounts of energy quanta; the sharp line of the atomic absorption spectrum then is replaced by a broad absorption band. In individual atoms absorption and emission take place at the same wavelength, while in a whole molecule the absorption and emission spectrum do not coincide; the peak of the emission spectrum is at a longer wavelength than the peak of the corresponding absorption spectrum (Fig. 4.2a).

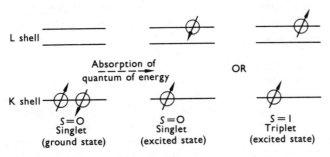

Fig. 4.1   Energy levels of electrons in the helium atom.

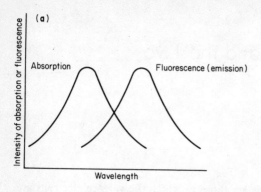

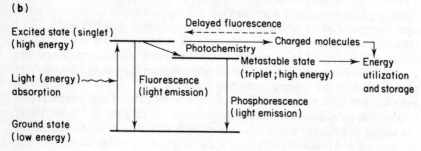

Fig. 4.2 (a) Spectra of light absorption and emission, (b) Energy level diagram.

4.1   Time spans, fluorescence, phosphorescence and luminescence

A molecule, say of chlorophyll, in an electronically excited state can revert to the ground state in a number of ways. It can transfer the electronic excitation to another acceptor molecule as is the case in photosynthesis. It can also dissipate part of the acquired energy as heat and emit a photon back into space (Fig. 4.2b). This phenomenon is called *fluorescence*. The wavelengths of a fluorescence spectrum are always longer than the wavelengths of the corresponding absorption spectrum. Chlorophyll *a* extracts, for example, absorb in blue and red regions of the spectrum but fluoresce only in the red; maximum intensity of fluorescence occurs at 668 nm compared to the wavelength of maximum absorption at 663 nm. The average time a molecule has to spend in the excited state between absorption and emission is known as the natural lifetime of the excited state; its duration depends on the electronic properties of the excited state. The average fluorescence lifetime is of the order of 10^{-9} s.

   A third route by which an excited molecule can lose its energy is by transfer from its original excited singlet state into a metastable triplet state with a much longer lifetime (of the order of milliseconds) by a mechanism called intersystem crossing. From the metastable triplet state the molecule can revert to the natural ground state by emitting a photon at a longer wavelength. This weak emission is known as *phosphorescence*. Phosphorescence is slow enough to be observed by the eye even when the exciting light is turned off. Due to their longer lifetime, lower energy, and magnetic moment (since the excited electron and its partner have parallel spins) the *triplet excited states* can be of importance in photochemistry. As yet there is no conclusive evidence for the participation of chlorophyll triplet states in photosynthesis although they are useful as a diagnostic tool.

The primary event in all photosynthetic processes is the light-induced transfer of an excited electron from a donor species D to a closely bound acceptor A. In chloroplasts, D is a special type of chlorophyll conjugated to a protein and located in the *Reaction Centre* (§4.4). Thus:

$$DA \xrightarrow{\ h\nu\ } D^{*}A \longrightarrow D^{+}A^{-}$$

where D^{*} is a chlorophyll molecule which has acquired an exciton (quantum of excited energy). The energy trapped in the charge separation is subsequently utilized for photosynthetic electron transport – thus light energy is converted to chemical energy. Antenna pigments transfer their absorbed electronic excitations to the reaction centre as singlet excitons and the reaction centre photochemistry is also initiated from an excited singlet state. The initial charge separation event happens within a time span of a few pico (10^{-12}) seconds after the capture of a photon.

When chloroplasts are illuminated at room temperature and immediately kept in the dark they are able to emit light, the emission resembling that of chlorophyll fluorescence. This light emission, observable from microseconds to minutes after terminating illumination, is called *delayed fluorescence* or *delayed light emission*. Delayed fluorescence is caused by the spontaneous recombination of some of the charged species formed in the reaction centre ($D^{+}A^{-}$) to regenerate D^{*}, and release of chemical energy as light. The intensity of delayed fluorescence is a reflection of the energized state (proton gradient) of the thylakoid membranes and of the rate of electron transfer from A^{-} to subsequent components of the electron transport chain. Delayed light emission can also occur by a reversal of the energy transition from the excited singlet state to the excited triplet state (Fig. 4.2b). The exact mechanism of the process is still not fully understood.

If, on the other hand, chloroplasts are frozen rapidly after illumination (or illuminated at low temperature, e.g. $-196°C$, liquid N_2) and subsequently warmed in the dark sudden bursts of light emission (glow) are observed at specific temperatures. This phenomenon is referred to as *thermoluminescence*, which is a special type of delayed light emission. In frozen chloroplasts the metastable charged species generated by light absorption are stabilized, but on thermal activation they recombine and release energy as light.

4.2 Energy transfer or sensitized fluorescence

The phenomenon of sensitized fluorescence involves the interaction of two molecules that may be separated in solution by many molecules of the solvent. In this type of energy transfer, two kinds of pigment are dissolved in the same solvent and the solution is illuminated with light of such wavelength that can be absorbed by only one of the pigments called the *donor*. The wavelength of light emitted from the solution, however, corresponds to the fluorescence spectrum of the second pigment molecule, the *acceptor*. The energy of excitation of the donor molecule is transferred by resonance to the acceptor molecule. One of the requisites for this type of energy transfer is that the fluorescent state of the donor molecule must have an energy greater than or equal to the fluorescent state of the acceptor molecule, that is, the fluorescence band of the donor molecule should overlap with the absorption band of the acceptor (Fig. 4.3). The quanta taken up by accessory pigments in many blue-green algae are transferred either wholly or partly to chlorophyll *a* by sensitized fluorescence. In the green algae chlorophyll *a* fluorescence is observed when light is absorbed by chlorophylls *a* and *b* and also by the carotenoids. Chlorophyll *a* has its fluorescence at the longest wavelength so that the migration of energy is always from the other excited chloroplast pigments to chlorophyll *a*.

From studies of chlorophyll *a* fluorescence and quantum yields of photosynthesis in *Chlorella* it was shown that in this alga the transfer of excitation energy from chlorophyll *b* to *a* is 100% efficient while the energy transfer from the carotenoids to chlorophyll *a* is only 40% efficient. The close assembly of various pigment molecules in the lamellae is necessary for efficient energy transfer by this type of inductive resonance.

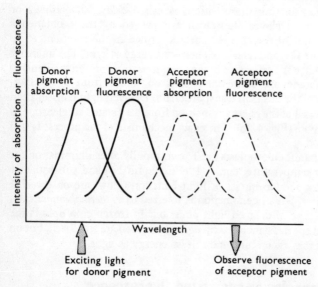

Fig. 4.3   Relationship between absorption and fluorescence spectra of donor and acceptor pigments.

4.3 Emerson effect and the two-light reactions

A plot showing the efficiency of photosynthesis (measured as O_2 evolution or CO_2 fixation) by monochromatic light as a function of the wavelength of light is known as the *action spectrum* of photosynthesis. For photochemical reactions involving a single pigment the action spectrum has the same general shape as the absorption spectrum of the pigment. If P molecules of O_2 per second are evolved from a system which absorbs I quanta of monochromatic radiation per second, then the ratio P/I, (Φ), is called the quantum yield or *quantum efficiency* of photosynthesis. The reciprocal of the quantum yield $[1/\Phi]$ which gives the number of quanta required to liberate one molecule of O_2 is usually called the *quantum requirement* of photosynthesis. Although values ranging from 4 to 12 have been mentioned for the quantum requirement by various workers in the past the widely accepted value is a minimum of 8.

Emerson and associates at the University of Illinois in the 1940s studied the action spectra of photosynthesis for various algae by measuring the maximum quantum yield of photosynthesis as a function of the monochromatic light used to illuminate the algae. They found that the most effective light for photosynthesis, in *Chlorella*, was red (650 to 680 nm) and blue (400 to 460 nm), those colours that are most strongly absorbed by chlorophyll. The photosynthetic efficiency of a quantum absorbed at 680 nm was about 36% more than that of a quantum absorbed at 460 nm.

The quantum yield of photosynthesis decreased very dramatically with increasing wavelength beyond 685 nm even though chlorophylls still absorb light at these wavelengths. This fact, the so-called *red drop* in photosynthesis, could not be explained at that time. However, Emerson and co-workers later showed that the amount of photosynthesis in far red light (wavelengths greater than 685 nm) could be increased considerably by a supplementary beam of red

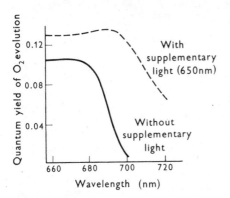

Fig. 4.4 Efficiency of photosynthesis (quantum yield) in the green alga *Chlorella* at different wavelengths of light. On adding supplementary light the quantum yield is enhanced at wavelengths above 680 nm – the Emerson enhancement effect. (Redrawn from Emerson et al., (1957). *Proceedings of the National Academy of Science, U.S.*, **43**, 133.)

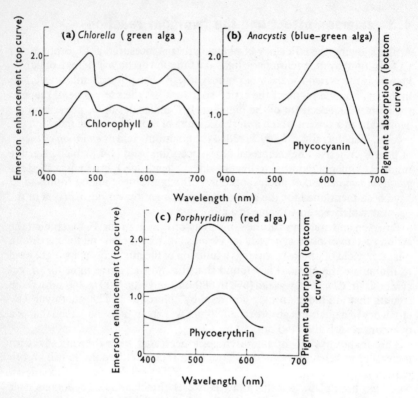

Fig. 4.5 Action spectrum of the Emerson effect in different algae (upper curve in each case) correlated with the absorption of the accessory pigments of the same algae (lower curve). (a) *Chlorella* containing chlorophyll *b*; (b) *Anacystis* containing phycocyanin; (c) *Porphyridium* containing phycoerythrin. (Redrawn from Emerson and Rabinowitch (1960). *Plant Physiology*, **35**, 477.)

light (about 650 nm) (see Fig. 4.4). In fact the total amount of photosynthesis carried out in the presence of a mixture of far red light and red light was greater than the sum of the amounts of photosynthesis carried out in separate experiments with the individual beams of light. This increase of the photosynthetic efficiency of far red light in the presence of a supplementary beam of lower wavelength is known as the *Emerson enhancement effect*. Experimental results of this effect are shown for three algae with different accessory pigments in Fig. 4.5. What is actually measured is O_2 evolution. From studies on the action spectrum of photosynthesis and of chlorophyll fluorescence Emerson and Rabinowitch concluded that photosynthesis is enhanced by energy absorption by accessory pigments and transfer of this absorbed energy to chlorophyll *a* by the process of 'inductive resonance'. To explain the Emerson enhancement effect they put forward the hypothesis that at least two photochemical acts are involved in photosynthesis and are preferentially sensitized by different pigments and that each electron must be photoactivated *twice* on its path from the primary donor, water, to the ultimate acceptor, CO_2. They also postulated the existence of two types of chlorophyll *a* in the cell, one associated with a reductant and another associated with an oxidant, and one of these is closer to the accessory pigment than the other – these are now termed photosystem I and photosystem II respectively.

Important subsequent work by Myers and French in 1960 showed that photosystem I and photosystem II need not be excited simultaneously to get optimum photosynthesis but that they may be illuminated alternately with a short dark period of a few seconds between them. This indicated that the two-light reactions could store their photochemical products for a short time before reacting with the electron transfer chain.

Experimental evidence and theoretical postulates in support of the two-light reaction hypothesis followed. The difference in electrode potential (ΔE) between the reactants and the final products in the photosynthetic reaction is 1.25 V (ΔE of CO_2–glucose couple = – 0.43 V; ΔE of H_2O–O_2 couple = + 0.82 V).

Photosynthetic reduction of CO_2 can be summarized by the equations:

$$2H_2O \xrightarrow{\text{light}} O_2 + 4H^+ + 4e^-$$

$$CO_2 + 4H^+ + 4e^- \longrightarrow (CH_2O) + H_2O$$

Thus four electrons are required to be transferred from water, through a redox span of 1.25 V, to reduce one molecule of CO_2. The energy required for the reduction of one mole of CO_2 is $1.25 \times 4 \times 9.64 \times 10^4 = 48.2 \times 10^4$ J ($\S1.6$). Theoretically this energy requirement can be satisfied by the capture of four mole quanta (one photon per electron) of photosynthetically active red light,

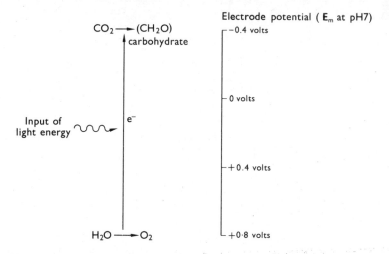

Fig. 4.6 Redox potentials of the overall reaction of photosynthesis.

say 700 nm, which have an energy content of $4 \times 17.1 \times 10^4 = 68.4 \times 10^4$ J. However, due to thermodynamic losses during energy conversion, only a fraction of the absorbed photon energy is converted to chemical free energy – in chloroplast photosynthesis this fraction seldom exceeds 0.36. Thus the maximum chemical free energy available for photosynthesis from a mole quantum of 700 nm light is $17.1 \times 10^4 \times 0.36 = 6.156 \times 10^4$ J. The minimum number of quanta required to reduce a molecule of CO_2 is therefore $48.2 \times 10^4/6.156 \times 10^4 = 8$. If we accept this value of 8 as the quantum requirement for photosynthesis then, since 8 quanta are consumed in the transport of 4 electrons (for one O_2 evolved) and since each quantum can activate only one electron at a time, it is logical to infer that each of the four electrons has to be activated by two separate light reactions. In 1960 Hill and Bendall at Cambridge put forward the idea that the two-light reactions should be in series (and not in parallel) with cytochromes b_6 and f acting as electron carriers in the 'dark' reaction which connects the two photosystems. This is shown in Fig. 4.7 and is elaborated and updated in Chapter 5. The most conclusive evidence for two separate photosystems came from a series of difference spectroscopy studies initiated by Duysens and by Kok and elaborated further by Witt. In this type of measurement the change in absorbances of various constituents of a cell suspension are studied by illumination with mono-chromatic light of varying wavelengths. A schematic diagram of a difference spectrophotometer is shown in Fig. 4.8. By following the qualitative and quantitative changes in the difference spectrum of individual photosynthetic reaction components, e.g. cytochromes, the role of some of these components in the electron transfer pathway can be inferred. For example, Duysens illuminated a suspension of the red alga *Porphyridium* in the presence of DCMU (a synthetic weed killer which inhibits oxygen evolution) and found an accumulation of cytochrome f in the oxidized

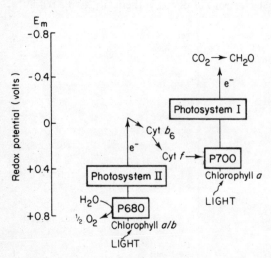

Fig. 4.7 The two-light reactions of the 'Z' scheme of photosynthesis. (Concept of Hill and Bendall, (1960). *Nature*, **186**, 136.)

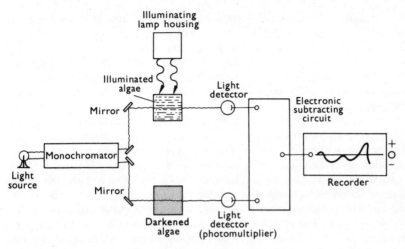

Fig. 4.8 Diagram of a difference spectrophotometer to measure the absorption of pigments.

form. When the experiment was repeated with unpoisoned alga there was no change in the cytochrome f spectrum. Duysens concluded that cytochrome f was an intermediate in the reactions accompanying oxygen evolution (see also Chapter 5).

4.4 Reaction centres and primary electron acceptors

By illuminating suspensions of algae with brief flashes of light Kok was able to identify and characterize a special type of chlorophyll *a* which he called P700. P700 is a trace constituent of chloroplasts with an absorption peak around 700 nm. It is reversibly bleached in light; the bleaching corresponds to an oxidation of P700 to P700$^+$. Subsequent investigations have shown that P700 is the reaction centre 'trap' pigment in which all incident light energy of wavelength greater than about 680 nm is captured and utilized for primary photochemical reactions, i.e. P700 is the primary electron donor of PSI. P700 is a special type of chlorophyll *a* (Chl *a*); evidence available at present is not conclusive as to whether it is a monomer or a dimeric chlorophyll molecule. Recently it has been reported that P700 in spinach chloroplasts is a dimer of Chl *a'*, which is a C_{10} epimer (with positions of H and CH_3COO interchanged at C_{10}) of Chl *a* (Fig. 3.8). Light displaces an electron from excited P700$^+$ and simultaneously an electron moves from plastocyanin (or from a *c*-type cytochrome in some blue-green algae) to P700$^+$. The nature of the primary electron acceptor from P700$^+$ is still not definitely established. At one time (in the 1960s) it was thought to be ferredoxin. However, in the last 15 years a number of electron mediator species functioning between P700 and ferredoxin have been detected mainly by measuring optical absorption changes and electron paramagnetic resonance (EPR) signals which result from flash excitation of chloroplasts or purified PSI particles at very low temperatures (liquid N_2 and liquid He). Present evidence suggests that the primary acceptor of electrons from P700$^+$ is a special Chl a molecule designated as A_0. The electron is transferred from A_0 to ferredoxin via a series of membrane-bound electron mediators which are identified by the symbols A_1, X and Fe-S_A and Fe-S_B (see Table 3.2 and Fig. 5.2).

Photosystem II has a higher content of chlorophyll b than PSI and is strongly fluorescent. The reaction centre chlorophyll of PSII is designated as P680. Witt and co-workers followed chlorophyll absorption changes and rates of O_2 evolution in flashing light experiments in normal and DCMU-treated chloroplasts. They found a correlation between the O_2 evolution rate and the photo-induced absorption change of a chloroplast component with a peak around 682 nm. Further investigations have shown that P680 is also a specialized type of chlorophyll a. The primary electron acceptor from P680$^+$ in PSII has been identified as a pheophytin. From pheophytin the electron migrates, via two quinones Q_A and Q_B, to a plastoquinone pool which serves as an electron reservoir between the two photosystems.

4.5 Experimental separation of the two photosystems

In recent years various attempts have been made to separate physically PSI and PSII from chloroplasts. Chloroplasts can be fragmented by the addition of detergents like digitonin or sodium dodecyl sulphate or Triton X-100, by sonic vibration or by mechanical extrusion under high pressure (about 800 atmospheres) through a French or Yeda Press. The fragments are then separated by a combination of differential and density gradient centrifugation and chromatography techniques. Some of the preparations thus obtained had a very high chlorophyll a to chlorophyll b ratio and a high P700 content suggesting they are highly enriched in PSI constituents. Particles enriched in PSI or PSII constituents have also been isolated from blue-green algae. Plant and algal mutants lacking either PSI or PSII have been produced by genetic manipulation and this should help in the isolation of pure PS particles.

A PSII reaction centre core complex consisting of four chlorophyll and two pheophytin molecules and six polypeptides: 47kD, 43kD; two 32kD (D_1 and D_2) and the 9 and 4kD sub-units of cytochrome b559 was isolated from spinach grana thylakoids in 1987. This spinach PSII complex, like the purple bacterial reaction centre complex (section 7.3), is able to catalyse light-induced primary charge separation and electron transport.

By treatment with detergents the proteins tightly bound to the thylakoid membranes can be brought into solution. The soluble components can then be resolved into fractions of varying molecular weights by electrophoresis on gels, e.g. polyacrylamide (PAGE), and located as specific bands in the gel by staining with dyes. Also it is possible to select mutants of plants and algae lacking specific chloroplast pigments or proteins and often deficient in some key photosynthetic activity or spectral characteristic. By comparing the gel bands from chloroplasts of such mutant and wild type plants it is possible to assign a

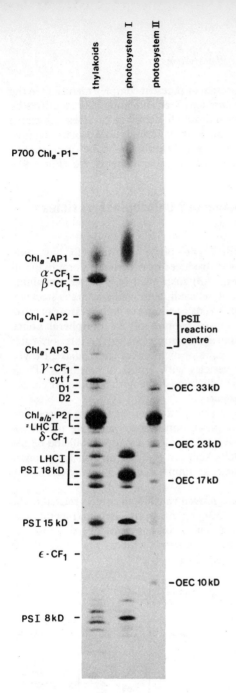

Fig. 4.9 Polypeptide patterns, obtained by polyacrylamide gel electrophoresis of barley thylakoids, PSI particles, and PSII particles. **AP**, apoprotein; **CF$_1$**, coupling factor extrinsic to the membrane; **D$_1$**, herbicide (Q$_B$) – binding protein; **D$_2$**, 32 kD (Q$_A^-$ binding) PSII protein of as yet unknown function; **LHC**, light-harvesting pigment–protein complex; **OEC**, oxygen-evolving complex. (Courtesy of Dr. D. Simpson and Professor D. von Wettstein, Carlsberg Laboratory, Copenhagen.)

definite function to an individual protein or pigment complex identified in the gel. Some of the thylakoid components separated and tentatively identified by the above techniques are: a P700-chlorophyll *a*-protein complex associated with the PSI reaction centre, a light harvesting chlorophyll *a/b* protein complex associated with the PSII reaction centre, membrane-bound iron-sulphur proteins, cytochrome *f*, ATPase-coupling factor components, etc. (Fig. 4.9).

4.6 Inside-out and right-side-out chloroplast vesicles

The technique of aqueous-polymer two phase partition for the sub-fractionation of chloroplasts has been introduced by Albertsson's group. In this method granal chloroplasts (see Fig. 3.3) suspended in a buffer containing 150 mM NaCl or 2 mM $MgCl_2$ (cationic medium to maintain grana stacking) are disintegrated by passing through a Yeda press at very high pressure of nitrogen. The Yeda press extrusion causes rupture of the peripheral granal membranes but the granal membranes stay appressed in the interior. After rupture, resealing of the appressed pairs of membranes from adjacent thylakoids results in the formation of vesicles with 'inverted' sidedness. These vesicles are called *inside-out vesicles* to distinguish them from normal thylakoid vesicles which are termed *right-side-out*.

The inside-out and right-side-out vesicles differ in their surface charges and are separated by aqueous polymer phase partition for example in a 5.7% dextran (MW ~ 500 000), 5.7% polyethylene glycol (MW ~ 3500) and 88.6% water system. The inside-out vesicles settle in the dextran-rich lower phase while the normal right-side-out vesicles accumulate in the polyethylene glycol-rich upper phase.

The vesicles thus fractionated have proved valuable in studies related to the topography and function of various components in the thylakoid membrane. Inside-out vesicles are rich in photosystem II components suggesting that this photosystem is located in the granal membranes whereas photosystem I is concentrated more in the agranal membranes.

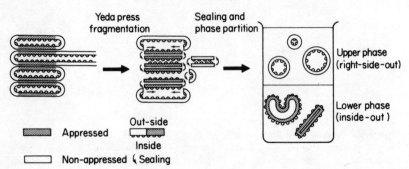

Fig. 4.10 Schematic illustration of the formation of right-side-out and inside-out membrane vesicles from thylakoids. (After Andersson, Sundby, Akerlund and Albertsson (1985). *Physiologia Plantarum* **65**, 322–30.)

4.7 Photosynthetic oxygen evolution

The water-splitting (oxygen-evolving) reaction:

$$2H_2O \longrightarrow O_2 + 4H^+ + 4e^-$$

is carried out on the internal surface of the thylakoid membrane on the oxidizing side of PSII (Fig. 5.2). Each quantum captured by the PSII reaction centre oxidizes one P680 molecule and generates one $P680^+$ Pheo$^-$ complex (Pheo = pheophytin). $P680^+$ is reduced to P680 by the removal of an electron from Z, the primary donor of PSII; Z probably is a tyrosyl residue. The oxidized species Z^+ immediately accepts an electron from the oxygen-evolving complex (designated M) which ultimately removes electrons from water so releasing oxygen. The events can be schematically represented by the equations:

$$\text{PSII (P680--Pheo)} \xrightarrow{h\nu} P680^+ \text{ Pheo}^- \tag{1}$$

$$P680^+ + Z \longrightarrow P680 + Z^+ \tag{2}$$

$$Z^+ + M \longrightarrow Z + M^+ \tag{3}$$

$$M^{4+} + 2H_2O \longrightarrow M + O_2 + 4H^+ \tag{4}$$

Notice that four positive charges need to be accumulated on M before an oxygen molecule can be released.

Kok and Joliot measured oxygen evolution from dark-adapted chloroplasts and algae illuminated by a sequence of saturating light flashes. They observed that the maximum yield of oxygen per flash was obtained, initially, after the third flash and subsequently after every fourth flash, i.e. oxygen peaks occurred at flashes 3, 7, 11, etc. To explain these periodic oxygen peaks Kok and Joliot proposed the '*S state hypothesis*' – the oxygen-evolving complex M can cycle through five different oxidation states, viz. S_0, S_1, S_2, S_3 and S_4, each state differing from the preceding one by the loss of a single electron (Fig. 4.11).

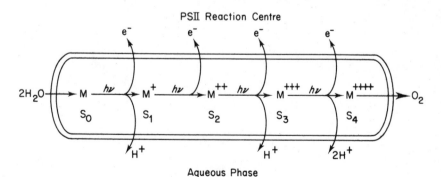

Fig. 4.11 Schematic illustration of the formation of S states and ejection of protons during photosynthetic water-oxidation resulting in O_2 evolution.

When four electrons are lost by M, i.e. when M has reached the S_4 state, it is able to react with water producing oxygen and returning to the uncharged state, S_0. Dark adaptation of chloroplasts synchronizes most of the oxygen-evolving centres to the S_1 state – hence maximum oxygen evolution occurs after the third flash at the start of the flash regime. During water oxidation protons are released into the interior of the thylakoid vesicles, the sequence of proton release probably occurring as shown in Fig. 4.11.

The structure of the oxygen-evolving complex M and the exact mechanism of water oxidation are still not clearly understood. Studies using inhibitors of oxygen evolution, spectroscopic data, and reconstitution experiments suggest that M may be a mangano-protein complex containing four atoms of Mn, two of which are essential for water oxidation. Reagents such as NH_2OH and NH_2NH_2 at low concentrations reduce the S_1 centres and inhibit oxygen evolution. Heat treatment of chloroplasts (55°C for 5 min) and washing in 1M Tris, 1M divalent salts, or in alkaline buffer (pH 8) dissociate the proteins of the complex M, release Mn, and inhibit oxygen evolution. Depletion of chloroplasts of chloride ions or in some instances of calcium ions also inhibits oxygen evolution – these ions may play a physiological role in maintaining the integrity of M.

5

Photosynthetic Electron Transport and Phosphorylation

In Chapter 2 we elucidated the idea of photosynthesis involving both light and dark phases in the fixation of CO_2. The recognition and experimental demonstration of these two phases was an important step towards the modern understanding of the CO_2 fixation process. This was not possible until Arnon, Allen and Whatley in 1954 were able to isolate chloroplasts from spinach leaves which were capable of carrying out complete photosynthesis, i.e. fixing CO_2 to the level of carbohydrate (see Chapter 3 for the preparation technique of chloroplasts). They were able to physically separate the light and dark phases and to show the light-dependent formation of ATP and $NADPH_2$ which then acted as the energy sources for the subsequent dark fixation of CO_2. This is summarized in the familiar diagram below (Fig. 5.1).

The light phase, which occurs subsequent to the initial light reactions discussed in the previous chapter, involves biochemical reactions with life times of 10^{-2} to 10^{-5} s. The initial light reactions have of course much shorter lifetimes – down to 10^{-15} s. The biochemical events of the light phase result in (*i*) the production of the strong reducing agent, $NADPH_2$, (*ii*) the accompanying evolution of O_2 as a by-product of the splitting of H_2O and (*iii*) the formation of ATP which is coupled to the flow of electrons from H_2O to NADP.

In this chapter we shall discuss how these reactions are thought to occur, what evidence there is for making such assumptions and what compounds are involved in the sequence of electron transport reactions.

5.1 Reduction and oxidation of electron carriers

The production of $NADPH_2$, ATP and O_2 in the lamellae of the chloroplast involves the transfer of electrons through a chain of electron carriers. This

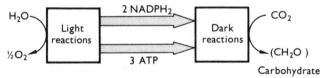

Fig. 5.1 Major products of the light and dark reactions of photosynthesis.

electron transfer requires that each of the carriers in turn becomes reduced and oxidized in order that the energy in the electron can be passed along the chain. Reduction simply means the adding of an electron while oxidation implies the removal of an electron from a compound. Whenever an electron is exchanged between two compounds one is oxidized and the other reduced. Almost every such exchange is accompanied by the release or absorption of energy. It makes no difference whether we think of the energy as arising out of the pull exerted on the electron by 'oxidizing power' or the push exerted by 'reducing power'.

Often, though not invariably, an electron travels in company with a proton, i.e. as part of a hydrogen atom. In that case oxidation means removing hydrogen and reduction means adding hydrogen. Thus NADP is reduced to $NADPH_2$ and CO_2 to carbohydrate, by the addition of hydrogen atoms.

The oxidation-reduction ('redox') potentials of biological electron carriers are expressed on a voltage scale at biological pH's which indicates that the $H_2O \rightarrow O_2$ couple is very oxidizing, with a positive midpoint potential of $+0.82$ V while the $H^+ \rightarrow H_2$ (gas) couple is very reducing with a negative potential of -0.42 V. Most biological electron transfer reactions occur between these two extremes (see Table 5.1). We shall see later that, in fact, the process of photosynthetic electron transport takes place at between $+0.8$ V and -0.4 V. In order to reach these extremes of redox potential light energy is required.

5.2   Two types of photosynthetic phosphorylation

Photosynthetic phosphorylation is the production of ATP in the chloroplast by light-activated reactions; it can take place via two systems, non-cyclic and cyclic. In non-cyclic photophosphorylation ATP is generated in an 'open'

Table 5.1 Midpoint redox potentials [E_m] of some chloroplast components and reactions.

Chloroplast component or reaction	E_m (volts)
P680 of PSII	$+0.9$ (or more $+$ ve)
H_2O/O_2	$+0.82$
P700 of PSI	$+0.48$
Cytochrome f	$+0.35$
Platocyanin	$+0.36$
Fe-S centre R	$+0.29$
Platoquinone	0.0
Cytochrome b_6	-0.05 and -0.17
$NADPH_2$	-0.34
H^+/H_2	-0.42
Ferredoxin	-0.42
CO_2/CH_2O	-0.43
Fe-S centre A of PSI	-0.55
Fe-S centre B of PSI	-0.59
e^- acceptor (X) of PSI	-0.71

electron transfer system together with the evolution of O_2 from H_2O and the formation of $NADPH_2$ from NADP. In cyclic photophosphorylation the electrons cycle in a 'closed' system through the phosphorylation sites and ATP is the only product formed. These two systems are shown in Fig. 5.2a. The occurrence of cyclic photophosphorylation *in vivo* is still open to question.

5.3 Non-cyclic electron transport and phosphorylation

This is the light-requiring process in which electrons are removed from H_2O, resulting in the evolution of O_2 as a by-product (Hill reaction, Chapter 2), and the transfer of these electrons via a number of carriers to produce a strong, negative reducing potential with the subsequent formation of $NADPH_2$, a reducing agent with a potential of -0.34 V.
This is simply expressed as:

$$ NADP + H_2O \xrightarrow[\text{chloroplasts}]{\text{light}} NADPH_2 + \tfrac{1}{2}O_2 $$

The carriers which have been identified include chlorophyll *a*, pheophytin, quinones, cytochromes *b* and *f*, Fe-S centres, plastocyanin, reductase enzymes and ferredoxin.

ATP formation accompanies the transfer of electrons and protons via the quinones and involves side reactions which are obligatorily coupled to the electron transfer process. Thus $NADPH_2$ and ATP formation occur in the process of non-cyclic electron transport where electrons are removed from H_2O and donated to NADP with the coupled formation of ATP. This overall reaction can be expressed as:

$$ NADP + H_2O + 2ADP + 2Pi \xrightarrow[\text{chloroplasts}]{\text{light (2e}^-\text{)}} NADPH_2 + 2ATP + \tfrac{1}{2}O_2 $$

This equation implies that each H_2O is split in the chloroplast membrane under the influence of light to give off $\tfrac{1}{2}O_2$ molecule (an atom of oxygen) and that the two electrons so freed are then transferred to NADP, along with H^+s (protons), to produce the strong reducing agent, $NADPH_2$. Two molecules of ATP can be simultaneously formed from two ADP and two Pi (inorganic phosphate) so that energy is stored in the form of this high energy compound. The precise number of ATP molecules formed is unclear (p. 58).

The $NADPH_2$ and ATP are the 'assimilatory power' required to reduce CO_2 to carbohydrate in the dark phase, which will be discussed in the next chapter. Thus 'assimilatory power' represents the initial products of the conversion of light energy into chemical energy.

A diagrammatic representation of the electron flow pattern in non-cyclic phosphorylation is given in Fig. 5.2a. This formulation is derived from the elegant hypothesis of Hill and Bendall in 1960. The scale on the left shows very clearly the potential of all the electron carriers in the chain and implies that the

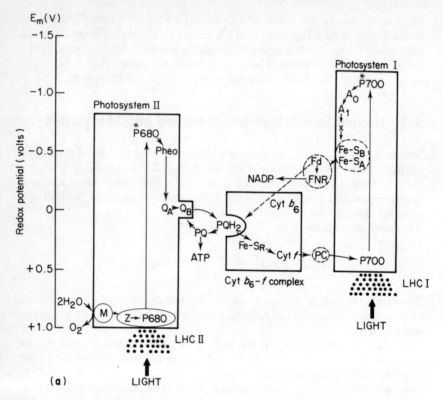

Fig. 5.2 (a) Electron transport scheme in chloroplasts. Rectangular boxes represent membrane-bound complexes. LHC I and LHC II, light-harvesting complexes of PSI and PSII respectively; other symbols as in Table 3.2.

sequence of electron flow depends to a large extent on their potential. This is a very neat view of electron transport and all the evidence to date indicates that it is most probably correct. The components of the non-cyclic electron transport pathway are organized into three complexes which span the chloroplast membrane. These complexes are the PSII complex, Cyt *b/f* complex and the PSI complex and can be separated from the membrane by detergent treatment. Plastoquinone, plastocyanin and ferredoxin are the mobile carriers which shuttle electrons across the complexes. The orientation of the electron carriers within the membrane itself is presently envisaged as in Fig. 5.2b. Note that the molecular weights are for polypeptides of barley thylakoids resolved by polyacrylamide gel electrophoresis and the values may differ for other species, eg. Mr of RC protein from spinach is 47kD.

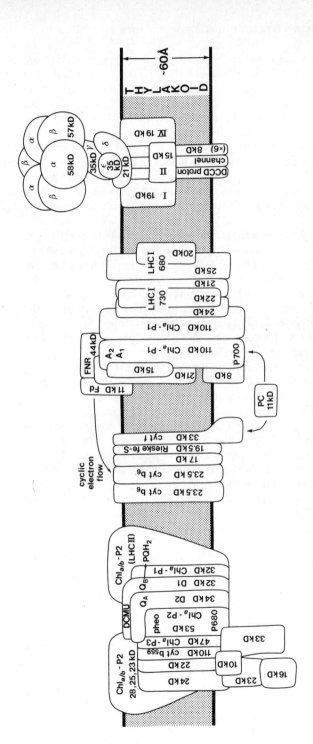

Fig. 5.2 (b) Schematic arrangement of polypeptides which have been identified as components of barley thylakoid membranes from freeze-fracture electron microscopy, gel electrophoresis, and biochemical studies. **EF$_s$**, endoplasmic fracture face of stacked thylakoids; **ES$_s$**, endoplasmic surface of stacked thylakoids; **PF$_s$**, protoplasmic fracture face of stacked thylakoids; **PF$_u$**, protoplasmic fracture face of unstacked thylakoids; **PS$_u$**, protoplasmic surface of unstacked thylakoids. The molecular weights of some peptides may vary from species to species. (Courtesy of Dr. D. Simpson and Professor D. von Wettstein, Carlsberg Laboratory, Copenhagen.)

What is immediately apparent is that two different light reactions are required to raise the electrons from the level of H_2O (+ 0.82 V) to the level of $NADPH_2$ (− 0.34 V) – these are designated photosystems II and I. Each of the systems has a different type of chlorophyll as the main light-absorbing pigment. Photosystem I has a predominance of chlorophyll *a* with an absorption maximum of 680 nm (bluish-green in colour) while photosystem II also contains the closely related chlorophyll *b* (see Chapter 3) which has its main absorption peak at 650 nm (yellowish-green in colour). Chlorophyll *a* occurs in all higher plants and algae, while chlorophyll *b*, an accessory pigment, is present in higher plants and green algae. Blue-green algae contain phycocyanin and allophycocyanins and red algae phycoerythrin as their accessory pigments for photosystem II.

The first evidence for the possible involvement of two light reactions in photosynthesis came from the work of Emerson and co-workers from 1943 on (see also Chapter 4). They showed in 1957 that the reduction of one molecule of CO_2 to carbohydrate required light of two different wavelengths – the quantum yield was very low if one light which was *only* absorbed by chlorophyll *a*, e.g. light of wavelength greater than 680 nm was used. The wavelength of the second light required to give efficient photosynthesis was found to correspond to that of chlorophyll *b* in higher plants and green algae and to the other accessory pigments in red and blue-green algae.

Further evidence for the requirement of light of two different wavelengths in non-cyclic electron flow from H_2O to NADP came from experiments which measured the changes in oxidation and reduction state of the cytochromes in algal chloroplasts. In Fig. 5.2 it is seen that cytochrome *f* is an electron carrier intermediate between photosystems I and II. Using very sensitive spectro-photometric techniques (see Chapter 4) it is possible to measure the redox state of the cytochrome *f*, due to its specific absorption peaks at 422 and 550 nm, under illumination of light of different wavelengths being absorbed by the two photosystems. The experimental results are shown in Fig. 5.3. Cytochrome in the chloroplast is naturally in the reduced state in the dark and thus if the algae are illuminated with light of 680 nm wavelength, which is mainly absorbed by photosystem I, i.e. chlorophyll *a* in the red alga, *Porphyridium*, electrons are removed from cytochrome *f* and donated to ferredoxin and thence to NADP – thus the cytochrome *f* becomes oxidized. Then if light at 562 nm (absorbed by photosystem II, phycoerythrin in *Porphyridium*, but also to some extent by photosystem I) is applied to the chloroplast the cytochrome *f* becomes reduced since it accepts electrons from photosystem II. Note that complete reduction is not achieved since photosystem I is still functioning to a limited extent in removing the electrons from the cytochrome *f*. In the dark the cytochrome *f* returns to its normal reduced state seen at the start of the experiment. This type of experiment was initiated by Duysens in 1961 and has proved of great value in localizing the site of electron carriers in the chain.

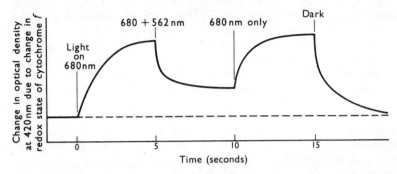

Fig. 5.3 Oxidation and reduction of cytochrome f in the red alga *Porphyridium*. Increase in OD420 is due to oxidation and decrease in OD420 is due to reduction of cytochrome f. Light of 680 nm is absorbed by photosystem I (chlorophyll *a*) and at 562 nm is absorbed by photosystem II (phycoerythrin). (After Duysens and Amesz (1962). *Biochimica Biophysica Acta*, **64**, 243).

Many different compounds have been isolated from chloroplasts (see Chapter 3) but the electron carriers shown in the scheme for non-cyclic electron transport (Fig. 5.2(a)) are those for which a role has been assigned so far. These compounds have been isolated and characterized chemically – quite a lot is known about ferredoxin and plastocyanin and now, recently, about cytochromes b_6 and f. Experiments have been devised which enable one to remove a specific carrier from the chloroplasts, e.g. by washing with water or low concentrations of detergents, organic solvent extraction or by mild sonication; a certain electron transfer reaction is thus diminished and then the re-addition of the extracted compound in its purified form restores the electron-carrying ability of the chloroplast membranes. This type of experiment has been successfully accomplished with plastoquinone, plastocyanin, ferredoxin, and the reductase enzyme (a flavoprotein) acting between ferredoxin and NADP.

In the last decade low temperature (-195 to $-269°C$) electron paramagnetic resonance spectroscopy has been applied successfully to detect a number of paramagnetic species as components of the chloroplast electron transport chain. This technique coupled with measurements of redox potentials under anaerobic conditions, has enabled investigators to assign locations for components such as pheophytin, the Fe–S centre R (Reiske) and Fe–S centres X, A and B, in the chloroplast electron transfer chain. All these iron-sulphur centres are bound to proteins located in the thylakoid membrane.

Elegant experiments by Levine, von Wettstein, Bishop, Ohad and others using genetic mutants of barley and the green algae *Chlamydomonas* and *Scenedesmus* have also helped to consolidate the electron flow sequence of Fig. 5.2. Specific mutants of the algae have been obtained which are deficient in certain parts of the electron chain, e.g. blocking electron flow between plastoquinone and cytochrome f or between plastoquinone and P700. With a knowledge of the exact site of the blockage experiments can be designed in which electrons are added and subtracted at various parts of the chain. This can be quite easily accomplished by using different types of dyes and is also used successfully in the spectrophotometric experiments mentioned earlier. These genetic and biochemical experiments which are similar in principle to those used with the mould *Neurospora* and the bacterium *Escherichia coli*, have given confirmatory evidence for the non-cyclic electron flow sequences and should prove valuable in our further elucidation of photophosphorylation and CO_2 fixation.

The use of chemical inhibitors of specific biochemical reactions is a classical and fruitful approach to understanding biochemical mechanisms. Non-cyclic photophosphorylation is no exception to the advantageous use of specific inhibitors. A number of different inhibitors have been found which block specific parts of the chain, e.g. DCMU [– (3,4-dichlorophenyl) – 1, 1-dimethyl urea] a herbicide which blocks oxygen evolution by binding to the Q_B protein of the PS II complex; antimycin A, an antibiotic, which prevents the reduction of cytochrome *f*; and KCN which inhibits plastocyanin (see §5.6).

Lastly we must discuss the ATP formation which is coupled to the non-cyclic electron flow. In 1958 Arnon, Whatley and Allen demonstrated the obligatory coupling of ATP formation to the reduction of NADP and showed that the rate of electron flow to NADP was dependent on the presence of ADP and Pi (required to form ATP). How many ATP molecules are formed per $NADPH_2$ produced is an important question because of the amount of ATP needed to fix CO_2 in the dark phase – 2 $NADPH_2$s and 3 ATPs are required per CO_2 molecule reduced to the level of carbohydrate. The present evidence suggests between 1 and 2 ATP are formed per $NADPH_2$.

The chemiosmotic hypothesis

Mitchell in 1961 proposed a '*chemiosmotic hypothesis*' to account for ATP formation accompanying electron transport in mitochondria and chloroplasts; for this he was awarded the Nobel prize in 1978. The chemiosmotic theory envisages that (*1*) intact membranes are almost impermeable to the passive flow of protons, (*2*) the electron donors and acceptors of electron and proton are arranged vectorially in the membrane, and (*3*) during photosynthetic electron transport protons are translocated from the external medium (stroma) into the osmotic space of the intrathylakoid membrane. Thus on illumination of chloroplasts for each quantum transferred to the reaction centre (P680 or P700) an electron is excited and transported through the membrane; simultaneously, a proton is taken up from the outside to the inside of the membrane. As a result of the movement of protons (and electrons) across the membrane, the membranes become energized and generate a transmembrane electrochemical potential referred to as the proton motive force (pmf). The magnitude of this proton motive force is expressed by the relation:

$$\text{pmf (volts)} = 2.303 \frac{RT}{F} \Delta pH + \Delta\psi$$

where R is the gas constant (8.314 J mol^{-1}/°K), F is the Faraday constant

(96.5 kJ mol^{-1} V^{-1}), T is temperature in °K, ΔpH is the transmembrane pH gradient, and $\Delta\psi$ the membrane electrical potential in volts. The energy available for ATP synthesis is equal to the product of the pmf and the number of protons consumed per molecule of ATP synthesized. According to the chemiosmotic hypothesis synthesis of ATP is promoted by the back-flow of protons via the enzyme, ATP-synthase (also called the coupling factor), located across the membrane. The energy released by the flow of protons is utilized for the synthesis of ATP, an endergonic reaction:

$$ADP + Pi \longrightarrow ATP \qquad \Delta G = 30 \text{ kJ (7.3 kcal)}$$

There are two sites in the thylakoid membrane where proton uptake accompanies electron transport. The reduction of plastoquinone by Q_B of PSII (Fig. 5.2) is thought to occur at the external surface of the membrane; for each electron accepted by a quinone in the plastoquinone pool one proton is taken up from the stroma. The second proton uptake reaction is the reduction of NADP$^+$ to NADPH in PSI catalysed by ferredoxin-NADP reductase, an enzyme located at the outer membrane surface. Protons are also released into the intrathylakoid space during electron transport by water oxidation and by the oxidation of plastoquinol by the cytochrome of the b–f complex (Fig. 5.2).

Data from several types of experiments support Mitchell's hypothesis. Firstly, if chloroplasts are illuminated under non-phosphorylating conditions in a medium depleted of ADP, they catalyse an electron transport-dependent proton uptake thereby generating a transmembrane proton and electron gradient both of which can be easily measured. The increase in intrathylakoid pH thus generated is of the order of 3 to 4 pH units; the thylakoid membranes can withstand a proton concentration gradient of 10 000:1. Secondly, if chloroplasts are preincubated in the dark in an appropriate buffer of low pH (pH = 4) and then rapidly injected into an alkaline buffer (pH = 8) containing ADP and Pi the synthesis of ATP occurs due to the establishment of a pH (proton) gradient. Although other theories to explain the phenomenon of ATP synthesis (photophosphorylation) have been put forward the chemiosmotic theory has gained wide acceptance among biochemists.

ATPase: coupling factors

The synthesis of ATP from ADP, Pi and protons is catalysed by the protein complex, ATPase or ATP synthase, also known as the coupling factor (CF) since it couples electron (and proton) flow to phosphorylation. In chloroplasts ATPase is located mainly in the non-granal (stromal) membrane (Fig. 5.2b, p. 55) and is composed of two distinctive components: factor CF_1, a hydrophilic protein projecting towards the stroma and CF_0, a hydrophobic protein which spans across the membrane. CF_1, which can be easily released from the membrane into aqueous buffers, has ATP hydrolase activity (after suitable treatment) and has a molecular weight of 400 kD. It is composed of five subunits named α, β, γ, δ, and ϵ in order of decreasing molecular weight. The functions of the subunits have been deduced from chemical modification and reconstitution studies. The δ subunit is required for binding CF_1 to CF_0, β (and probably α) is essential for catalytic activity, and the other subunits may play a regulatory role. CF_0 acts as a channel for pumping protons across the membrane to CF_1. Although several experimental techniques suggest that CF_1 undergoes conformational changes during proton and adenine nucleotide exchanges, the exact mechanism of ATP synthesis is still not established.

Uncouplers

Chemical reagents that stop ATP synthesis while allowing electron transport to proceed are known as *uncouplers*. These compounds allow the passive movement of protons across the membrane and restrict the formation of a proton gradient essential for phosphorylation. Chemicals such as carbonyl cyanide 3-chlorophenyl hydrazone (CCCP), and amines, make the membrane more permeable to protons. Weak bases, e.g. methylamine or ammonia in the neutral state, can readily diffuse into the thylakoids and there combine with H^+ ions; the protonated bases $CH_3NH_3^+$ and NH_4^+ cannot diffuse out of the membrane easily. Reagents such as gramicidin, dinitrophenol, valinomycin, and nigericin (in conjunction with K^+ ions) promote a cation–proton exchange across the membrane and thus dissipate the proton gradient. All these types of uncouplers have been used in studies to unravel the mechanism of photophosphorylation. ATP synthesis can also be blocked at the coupling site by inhibitors such as dicyclohexylcarbodiimide, triphenyltin, etc. and antibody to CF_1 which reacts directly with the coupling factors. These organic compounds are referred to as energy transfer inhibitors.

During the isolation and preparation of chloroplasts from the whole leaf the ATP formation factors are easily destroyed so that greater care must be taken in performing phosphorylation experiments than those in which only electron transport is measured. In Fig. 5.4 (pp. 62–3) a sequence of diagrams shows how one isolates chloroplasts and measures the O_2 evolution, $NADPH_2$ formation and ATP formation associated with non-cyclic photophosphorylation (see also Chapter 3).

5.4   Cyclic electron transport and phosphorylation

In this process, which requires light and chloroplasts, the only net product is ATP. The reaction was discovered in 1954 by Arnon, Allen and Whatley using isolated spinach chloroplasts and by Frenkel using chromatophores isolated from photosynthetic bacteria. It may be very simply represented by the following equation:

$$ADP + Pi \xrightarrow[\text{chloroplasts}]{\text{light}} ATP$$

Only a cyclic electron flow involving photosystem I is required in order to produce ATP. Figure 5.2a (p. 55) suggests how this may occur. Under the influence of an input of light an electron is removed from P700 in its excited state and donated to Fe–S centres and subsequently to ferredoxin which becomes reduced. The reduced ferredoxin then, instead of transferring its electron to NADP as in the case of non-cyclic electron flow, donates its electron to cytochrome b_6 and thence through the electron transport chain back to P700. Thus the electron undergoes a cyclic flow and the only measurable product is ATP which is formed by a coupling mechanism, probably similar to that involved in non-cyclic photophosphorylation even though the electron carriers may not be identical. The number of molecules of ATP formed per electron transferred is so far undetermined because of the difficulty in measuring the number of electrons cycling around the chain in a given time – the number of ATP's formed in a given time is, on the other hand, relatively simple to measure.

We can thus see that ferredoxin may play a central role in photosynthesis. It can donate electrons in a non-cyclic system to NADP in order to produce the strong reducing power in the form of $NADPH_2$ needed for CO_2 reduction; or it can donate electrons back into the electron transfer chain in a cyclic system resulting only in the formation of ATP. This ATP can be used for CO_2 fixation or for other reactions which only require ATP as their energy source, e.g. protein synthesis and the conversion of glucose to starch, both of which occur in the chloroplast. The physiological controlling mechanism associated with ferredoxin and its very low reducing potential of -0.42 V (equivalent to that of H_2 gas) have stimulated much research into the physicochemical and biochemical properties of this unique protein. A model for the active centre of ferredoxin is given in Fig. 5.5.

In the chloroplast ferredoxin is probably the physiological carrier (or cofactor) involved in cyclic photophosphorylation. However, experimentally we can replace ferredoxin by vitamin K_3, FMN or any of a number of dyes. The cyclic electron flow sequence may not be exactly the same as that for ferredoxin but again the only net product is ATP.

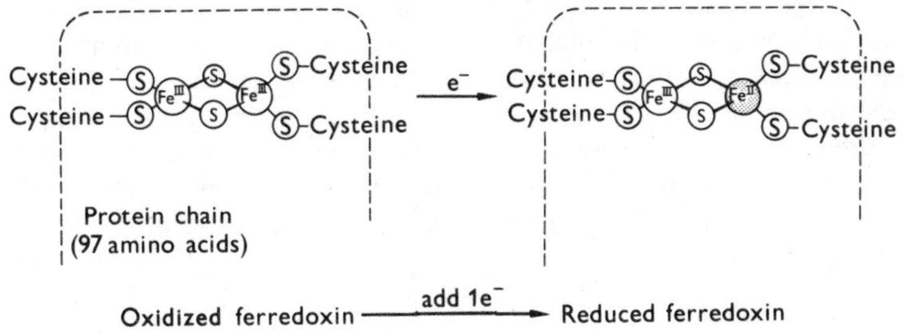

Oxidized ferredoxin $\xrightarrow{\text{add } 1e^-}$ Reduced ferredoxin

Fig. 5.5 A model for the active centre of plant ferredoxins showing the oxidized and reduced forms of the protein. (After Rao, Cammack, Hall and Johnson (1971). *Biochemical Journal*, **122**, 257.)

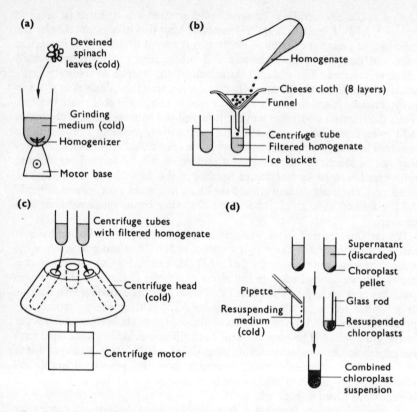

Fig. 5.4 Experiment to demonstrate photosynthetic O_2 evolution and $NADPH_2$ and ATP formation by isolated spinach chloroplasts.

It is also possible for the reduced ferredoxin to be oxidized by oxygen (rather than NADP) in the chloroplast under certain conditions. This non-cyclic electron transport from water to O_2 is designated as pseudocyclic electron transport. Such pseudocyclic electron transport results in O_2 uptake and formation of H_2O_2 ('Mehler reaction') which must be dissipated from the chloroplast since it can be toxic. Formation of ATP accompanies such pseudocyclic electron transport. The possible role of such an oxygen-catalysed pseudocyclic electron transport and phosphorylation in the regulation of CO_2 fixation and the protection of the chloroplast from damage by excess light is currently being investigated.

5.5 Structure — function relationships

The light phase of the overall CO_2 fixation to the level of carbohydrate has been shown to occur in the grana lamellae (or thylakoids) of the chloroplast while the

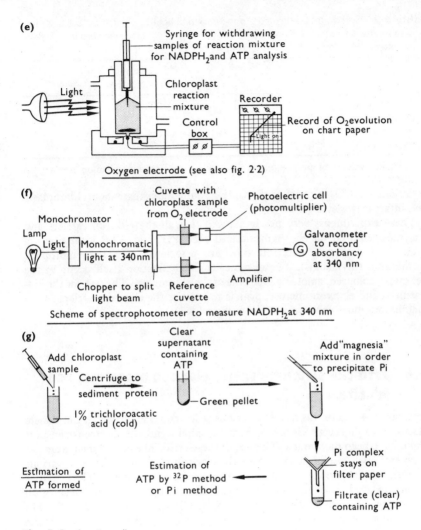

Fig. 5.4 (continued)

dark phase occurs in the stroma of the chloroplast. Arnon and co-workers demonstrated this in 1958 by physically separating the light and dark phase (see Table 5.2). Chloroplasts were illuminated in the absence of CO_2 and allowed to form large amounts of $NADPH_2$ and ATP with the concomitant O_2 evolution from the non-cyclic electron flow. The chloroplasts were then broken and the stroma separated from the grana lamellae which were discarded. Then in the dark, radioactive CO_2 was supplied and the enzymes in the stroma proceeded to

Table 5.2   Carbon dioxide fixation in the dark and light by chloroplast systems, i.e. stroma (yellowish matrix) and grana (chlorophyll-containing, green membranes) (Trebst, Tsujimoto and Arnon (1958). *Nature*, **182**, 351).

	$^{14}CO_2$ fixed (counts per minute)
Stroma (dark)	4 000
Stroma (dark) + grana (light)	96 000
Stroma (dark) + ATP	43 000
Stroma (dark + NADPH$_2$ + ATP	97 000

Note the equivalence of grana (light) and NADPH$_2$ + ATP, i.e. assimilatory power.

assimilate the CO_2 to produce the same carbohydrates that whole chloroplasts and intact leaves synthesize.

These experiments very neatly showed that all the electron carriers and enzymes required for the light-induced NADPH$_2$ and ATP formation via cyclic and non-cyclic electron flow are associated with the chloroplast membranes (thylakoids). The enzymes for CO_2 fixation itself occur in the yellowish-coloured, amorphous stroma of the chloroplast. The task for the biochemists and electron microscopists is to localize the electron carriers in the membranes more exactly (see Figs 5.2b and 8.1).

5.6   Artificial electron donors, electron acceptors, and inhibitors

In Chapter 4 and in this chapter, various redox compounds which can donate electrons to or accept electrons from the photosynthetic electron transport chain have been mentioned. The use of these artificial electron mediators has been extremely useful in measuring the electron transport activity of various segments of the photosynthetic pathway in isolated chloroplasts. The activities usually assayed are oxygen exchange (evolution or uptake), optical absorption change, or induction and decay of chlorophyll fluorescence. The specificity of an artificial mediator is dictated by its redox potential and its accessibility to the electron transport chain. Photosystem II has a redox span of + 800 mV (water oxidizing enzyme) to − 100 mV (reduced plastoquinone) and it has the potential to oxidize any redox molecule with an E_m more negative than 800 mV and to reduce any molecule with an E_m more positive than − 100 mV. Similarly, PSI can be expected to oxidize a mediator of E_m more negative than + 480 mV (E_m of P700) and reduce a mediator of E_m more positive than − 700 mV (E_m of centre X).

The accessibility of an electron mediator to its specific site depends on its solubility in lipid or aqueous phase, i.e. whether the molecule is hydrophobic or hydrophilic – some molecules may be amphiphilic. Hydrophobic molecules are more accessible to PSII components which lie embedded in the thylakoid membrane. On the other hand hydrophilic molecules are good electron acceptors from the reducing side of PSI which lies exposed on the outer side of the membrane.

Many electron mediators used to measure chloroplast electron transport are only stable either in the oxidized state or in the reduced state, and some of them inhibit electron transport when used in high concentrations. To overcome these drawbacks such mediators are usually added to the reaction medium in catalytic amounts in conjunction with a stable, non-inhibitory oxidant or reductant as the case may be. For example, the PSII electron acceptors para-phenylene diamine (PD), diaminodurene (DAD), and benzoquinone (BQ) are maintained in the oxidized state in reaction mixtures by the addition of excess ferricyanide. Similarly, when used as electron donors to plastocyanin (and PSI), the compounds N-tetramethyl-p-phenylene diamine (TMPD), dichloro-phenolindophenol (DPIP) or DAD are kept reduced by the presence of excess of ascorbate.

The accessibility of an electron donor or acceptor to the photosynthetic chain depends in addition on the structural integrity of the thylakoids in the chloroplast preparation. Thus, the hydrophilic compounds, ferricyanide and $DPIP_{ox}$ are used as PSI acceptors of more intact chloroplasts but will accept electrons also from PSII with broken chloroplasts.

Electron transport can be blocked at specific sites by the addition of compounds which bind to one of the components of the chain, remove one of the components, or alter the fluidity of the membrane structure. Antibodies raised against the proteins in the electron transfer chain specifically block by binding to their antigens. The more commonly used electron acceptors, donors, and electron transport inhibitors are shown in Fig. 5.6.

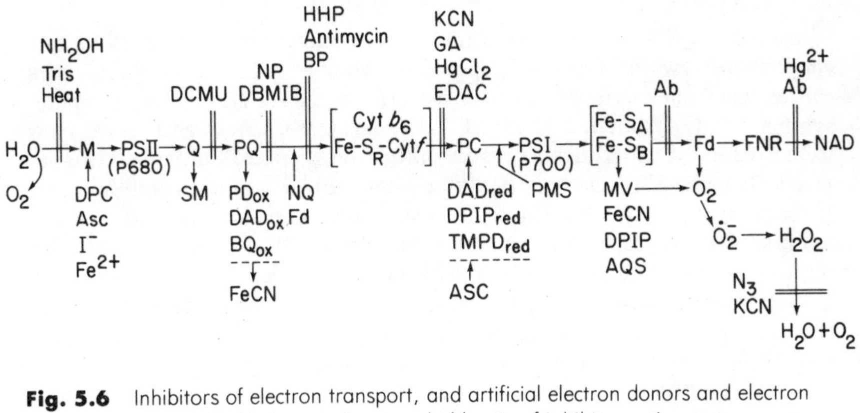

Fig. 5.6 Inhibitors of electron transport, and artificial electron donors and electron acceptors. Vertical double lines indicate probable site of inhibitor action; arrows towards the chain indicate site of electron donation; and arrows out of the chain indicate site of electron acceptance. **Ab**, antibody; **AQS**, anthraquinone sulphonate; **Asc**, ascorbate; **BP**, bathophenanthroline (iron-chelator); **BQ**, benzoquinone; **DAD**, diaminodurene (2,3,5,6-tetramethyl-p-phenylene diamine); **DBMIB**, 2,5-dibromo-3-methyl-6-isopropyl-p-benzoquinone; **DCMU**, 3-(3,4-dichlorophenyl)-1,1-dimethylurea; **DPC**, diphenylcarbazide; **DPIP**, dichlorophenol indophenol; **EDAC**, 1-ethyl-3-(3-dimethylaminopropyl)-carbodiimide; **FeCN**, ferricyanide; **GA**, glutaraldehyde; **HHP**, halogenated hydroxypyridine (cyclic electron transport inhibitor); **MV**, methylviologen; **NP**, nitrophenol; **NQ**, naphthaquinone; O_2^-, superoxide; **PMS**, phenazine methosulphate; **SM**, silicomolybdate; **TMPD**, N-tetramethyl-p-phenylene diamine.

6

Carbon Dioxide Fixation

In the previous chapter we have seen that $NADPH_2$ and ATP are produced in the light phase of photosynthesis. The fixation of CO_2 then takes place in the dark phase using the 'assimilatory power' of $NADPH_2$ and ATP. In this chapter we shall examine in some detail the reactions involved in the reduction of CO_2 to the level of carbohydrate since the reaction mechanisms and experimental techniques, so clearly worked out by Calvin and his co-workers from 1946 on, are some of the most important in modern biology. For his work in elucidating the path of carbon in photosynthesis Calvin received the Nobel Prize for Chemistry in 1961.

6.1 Experimental techniques

When the long-lived isotope of carbon, ^{14}C, became available in 1945 its use, coupled with two-dimensional paper chromatography developed a few years earlier, enabled experiments to be devised to investigate the pathway of photosynthetic $^{14}CO_2$ fixation. The unicellular green algae *Chlorella* and *Scenedesmus* were used in the experiments because of their biochemical similarity to higher green plants and because they could be grown under uniform conditions and subsequently very quickly killed in the short-time experiments used.

Three main types of experiments were performed to obtain the evidence required to postulate the detailed reactions of the cycle:

(a) Exposure of the photosynthesizing algae to $^{14}CO_2$ for different lengths of time. At the shortest times only the initial products will be radioactive. In this way phosphoglyceric acid (PGA) was identified as the primary carboxylation product; end-products such as sucrose became radioactive much more slowly.

(b) Determination of the position of radioactivity within the labelled compounds. In this way the details of the interconversions of sugar phosphate to regenerate the specific sugar phosphate which accepts the $^{14}CO_2$ molecule and the mechanism of synthesis of sugars and other compounds were worked out.

(c) Alteration of the external conditions, e.g. changing from light to dark, or changing from high to very low CO_2 concentrations, to see whether the cycle intermediates behave in a predictable manner.

The techniques employed are pictured in Figs 6.1, 6.2 and 6.3. Figure 6.1 is a diagram of the apparatus used for obtaining extracts of algae which have been

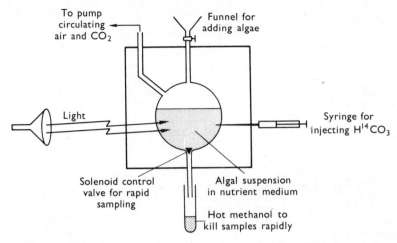

To pump circulating air and CO_2

Funnel for adding algae

Light

Syringe for injecting $H^{14}CO_3$

Solenoid control valve for rapid sampling

Algal suspension in nutrient medium

Hot methanol to kill samples rapidly

Fig. 6.1 Schematic representation of apparatus for studying $^{14}CO_2$ fixation in photosynthesizing algae.

photosynthesizing from $^{14}CO_2$. The algae are suspended in a nutrient medium through which air and CO_2 are bubbled with the pH of the whole suspension maintained constant. The control valve allows rapid removal of samples into methanol which immediately stops all reactions from proceeding further. Figure 6.2 illustrates the further handling of the inactivated algae. The killed algal extract is concentrated by vacuum and then applied directly to a chromatogram paper which is developed with different solvents in two directions at right angles to each other. The radioactivity of the compounds which are separated by this two-dimensional chromatography is measured – their location being known from radioautograms of the kind shown in Fig. 6.3. The two radioautograms in Fig. 6.3 show the compounds which contain ^{14}C in extracts of *Chlorella* which had been photosynthesizing for 5 and 15 seconds. It is seen that PGA, triosephosphate and sugar phosphates are formed very rapidly; sucrose, organic acids and amino acids are only formed after longer photosynthesizing times. The combination of radioactive CO_2 and two-dimensional chromatography is seen to be a very sensitive technique for detecting and quantitatively estimating the products of photosynthesis.

6.2 The photosynthetic carbon (Calvin) cycle

The fixation of CO_2 to the level of sugar (or other compounds) can be considered to occur in four distinct phases, as is shown in Figs 6.4 and 6.5.

I. *Carboxylation phase* This phase is thought to consist of a reaction whereby CO_2 is added to the 5-carbon sugar, ribulose bisphosphate, to form two molecules of PGA as follows:

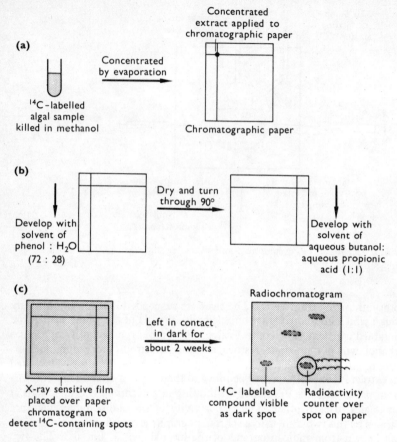

Fig. 6.2   Detection of the products of $^{14}CO_2$ fixation by algae after brief periods of illumination by the use of paper chromatography and autoradiography.

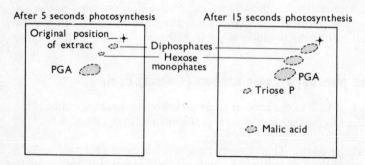

Fig. 6.3   Radioautograms of the photosynthetic products from $^{14}CO_2$ added for short periods of time to illuminate algae.

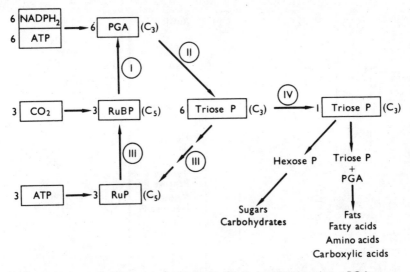

Fig. 6.4 Diagram of CO_2 fixation cycle. **RuBP** ribulose bisphosphate; **PGA**, phosphoglyceric acid; **Triose P**, phosphoglyceraldehyde; **RuP**, ribulose-5-phosphate.

$$\text{*CO}_2 + \underset{\substack{\text{Ribulose bisphosphate (RuBP)}}}{\begin{array}{c} CH_2OP \\ | \\ C = O \\ | \\ CHOH \\ | \\ CHOH \\ | \\ CH_2OP \end{array}} + H_2O \xrightarrow{\text{RuBisCO}} \underset{\substack{2 \times \textbf{Phosphoglyceric acid (PGA)}}}{\begin{array}{c} CH_2OP \\ | \\ CHOH \\ | \\ \text{*COOH} \end{array} + \begin{array}{c} CH_2OP \\ | \\ CHOH \\ | \\ COOH \end{array}}$$

This reaction is catalysed by the enzyme ribulose bisphosphate carboxylase (RuBisCO).

The evidence in support of this scheme is shown very clearly in Fig. 6.6. On illumination RuBP and PGA increase up to a certain level which is the so-called 'steady state' level in the photosynthesizing algae. When the light is switched off the RuBP content drops immediately (as light is needed for its synthesis) while the level of PGA rises – two molecules of PGA are formed from every molecule of RuBP which disappears. Also in Fig. 6.6 is shown the effect of changing from high to very low CO_2 concentrations. A 'steady state' level is achieved at 1% CO_2 but when the CO_2 level is suddenly decreased to 0.003% (with the light still on) the level of PGA drops quickly as there is insufficient CO_2 to fix; however, the level of RuBP increases since very little of it can be used to fix CO_2 into PGA, while it can still be formed in the light.

II. *Reduction phase* PGA formed by the addition of CO_2 to RuBP is essentially an organic acid and is not at the energetic level of a sugar. In order

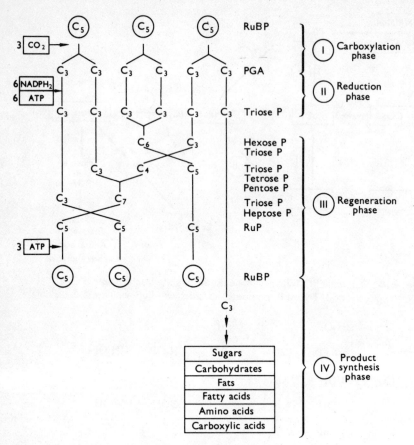

Fig. 6.5 Summary of reactions of photosynthetic CO_2 fixation.

for PGA to be converted to a 3-carbon sugar (triose P) the energy in the 'assimilatory power' of $NADPH_2$ and ATP must be used.

The reaction is in two steps: first, ATP-dependent phosphorylation of PGA at the COOH group to form 1,3-diphosphoglyceric acid and ADP, and second, reduction of the 1,3-diphosphoglyceric acid to phosphoglyceraldehyde by $NADPH_2$ releasing orthophosphate (Pi). The two steps can be summarized as follows:

$$
\begin{array}{l}
CH_2OP \\
| \\
CHOH + ATP + NADPH_2 \xrightarrow{\text{Enzymes}} \\
| \\
COOH
\end{array}
\qquad
\begin{array}{l}
CH_2OP \\
| \\
CHOH + ADP + Pi + NADP \\
| \\
CHO
\end{array}
$$

Phosphoglyceric acid   Phosphoglyceraldehyde
     (PGA)   (Triose P)

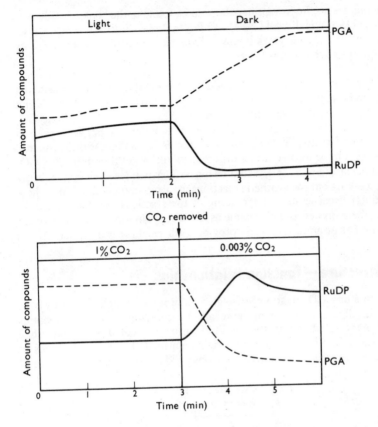

Fig. 6.6 Interconversions of RuBP and PGA, during experiments on photosynthesis.

It is seen that the reducing power of NADPH$_2$ is used to change the acid group of PGA to an aldehyde group of the triose P; ATP is required to provide the extra energy in order to accomplish this step but the Pi of ATP is not incorporated into the triose P. Both of the enzymes involved in the two steps have been shown to be present in isolated chloroplasts.

Once the CO$_2$ has been reduced to the level of the 3-carbon sugar, triose P, the energy-conserving part of photosynthesis has been accomplished. What is required thereafter is to regenerate the initial CO$_2$ acceptor molecule, i.e. ribulose bisphosphate, in order for the CO$_2$ fixation to continue again and again (regeneration phase) and to change the triose P to more complex sugars, carbohydrates, fats and amino acids (product synthesis phase).

III. *Regeneration phase* The RuBP is regenerated for further CO$_2$ fixation reactions by a complex series of reactions involving 3-, 4-, 5-, 6- and 7-carbon sugar phosphates which is depicted in summary form in Fig. 6.5. The details of the reactions are not important here but can be found in the books by Edwards and Walker (1983), Halliwell (1984) and Foyer (1984). Suffice it to say that all the reactions and the enzymes involved have been studied in some detail by various groups of research workers.

IV. *Product synthesis phase*   End-products of photosynthesis are considered primarily to be sugars and carbohydrates but fats, fatty acids, amino acids and organic acids have also been shown to be synthesized in photosynthetic CO_2 fixation. Many details of these synthesis reactions are known but again they do not concern us directly. What is, however, interesting is that the different end-products seem to be formed under different conditions of light intensity, CO_2 and O_2 concentration, as is depicted in Fig. 6.7. Much research is now being aimed at working out the synthetic reactions involved in the formation of these end products because an understanding of the reactions and the conditions favouring them may eventually enable us to induce plants to synthesize more or less sugars, fats or amino acids, by providing the required growth conditions.

6.3   Structure—function relationships

In order to study CO_2 fixation in isolated chloroplasts they must be isolated rather carefully in order to preserve all the components of the reactions involved. Arnon's laboratory showed in 1954 that this could be accomplished with all the CO_2 fixation products being identified. However, the rates of fixation were less than one tenth of those observed in leaves and it was not until

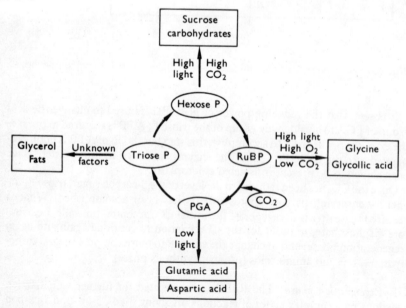

Fig. 6.7   Conditions favouring formation of secondary products in photosynthesis. (Redrawn from Hiller and Whatley, 1967.)

quite recently that Walker's and Bassham's laboratories were able, by using very careful isolation media and procedures, to obtain chloroplasts capable of CO_2 fixation rates approaching those in whole leaves. Again, the products of photosynthesis were the same as those observed by Calvin's group in whole algae and by Arnon's laboratory in the early isolated chloroplasts.

These studies emphasize that structural integrity is important in understanding how subcellular organelles such as chloroplasts do actually function. The way is now clear to study many more product synthesis reactions in isolated chloroplasts, e.g. starch and sucrose synthesis. Movement of inorganic phosphate and sugar phosphates in and out of the chloroplast is strictly controlled depending on metabolic requirements. However, basically we can say that the light phase occurs in the grana lamellae or membranes and that the dark phase occurs in the stroma or soluble part of the chloroplast.

6.4 Energetics of CO_2 fixation

If we look at the overall and constituent equations of photosynthesis again, we will be able to examine the energy conserving and energy expending parts of the carbon fixation cycle.

The general equation for formation of glucose can be represented by:

$$(i)\ CO_2 + H_2O \rightarrow [CH_2O] + O_2 \qquad \Delta G = +48 \times 10^4\ J\,(114\,kcal)$$

This means that 48×10^4 joules of energy are required to fix one mole of CO_2 to the level of glucose. This large positive value thus requires that a large amount of energy must be added.

We have already seen that this energy is derived from the light phase of photosynthesis and can be represented by the 'assimilatory power' of $NADPH_2$ and ATP. In order to fix one CO_2 molecule, two molecules of $NADPH_2$ and three of ATP are required (see Figs 6.4 and 6.5).

The energy present in the $NADPH_2$ and ATP can be represented as follows:

$$(ii)\ 2NADPH_2 + O_2 \rightarrow 2NADP + 2H_2O \qquad \Delta G = -44 \times 10^4\ J\,(-105\,kcal)$$

$$(iii)\ 3ATP + H_2O \rightarrow 3ADP + 3Pi \qquad \Delta G = -9.2 \times 10^4\ J\,(-22\,kcal)$$

This energy is sufficient to reduce one CO_2 molecule to the level of glucose with about 5×10^4 joules (13 kcal) to spare ($ii + iii - i$).

The constituent parts are:

	$\Delta G\,(J)$	$\Delta G\,(kcal)$
$(i)\ CO_2 + H_2O$ $\rightarrow [CH_2O] + O_2$	$+48 \times 10^4$	$+114$
$(ii)\ 2NADPH_2 + O_2$ $\rightarrow 2NADP + 2H_2O$	-44×10^4	$-105\ (2 \times 52.5)$
$(iii)\ 3ATP$ $\rightarrow 3ADP + 3Pi$	-9.2×10^4	$-22\ (3 \times 7.3)$
	$-5 \times 10^4\ J$ (excess)	$-13\ kcal$ (excess)

In summary we can present the equations as:

$$CO_2 + H_2O + 2NADPH_2 + 3ATP \longrightarrow [CH_2O] + O_2 + 2NADP + 3ADP + 3Pi$$

Thus we see that photosynthesis is essentially a reductive process since most $(44 \times 10^4/53.2 \times 10^4 = 83\%)$ of the energy required to fix a molecule of CO_2 is derived from the strong reducing agent, $NADPH_2$, which has a redox potential of -0.34 V. The redox potentials of sugars can be thought to be approximately -0.43 V; so the ATP is required to fix the CO_2 to this lower redox value.
  We know that the redox potential of $H_2O \rightarrow O_2$ is $+0.82$ V, so the overall change in redox potential is 1.25 V ($+0.82$ to -0.43 V). This can be converted into terms of energy using the equation:

$$\Delta G = -nF\Delta E$$
$$= -(4)(9.64 \times 10^4)(1.25) = -48.2 \times 10^4 \text{ J (116 kcal)}$$

where n = number of electrons (= 4 electrons per molecule of oxygen)
        F = the Faraday (= 9.64×10^4 J per volt equivalent)
        ΔE = difference in redox potential

It is seen that this ΔG value is very close to that for fixing one molecule of CO_2 $(48 \times 10^4 \text{ J} = 114$ kcal; equation (i) above) and shows quite nicely the inter-conversion of energy in terms of joules and redox potentials.
  Lastly, we can discuss the quantum efficiency of CO_2 fixation. Each mole quantum of red light at 680 nm contains 17.61×10^4 J of energy. Thus at least three $(48 \times 10^4/17.61 \times 10^4 = 2.7)$ mole quanta of 680 nm light will be required for one CO_2 molecule to be fixed. However, experimentally it is found that 8–10 quanta of absorbed light are required for each molecule of CO_2 fixed or O_2 evolved. From our knowledge of non-cyclic photosynthetic phosphory-lation we deduce that there are two different light reactions required to reduce NADP with the electrons from H_2O

$$2NADP + 2H_2O \xrightarrow[\substack{\text{2 light reactions} \\ \text{chloroplasts}}]{4e^-} 2NADPH_2 + O_2$$

Thus we need at least 8 quanta (4 quanta per 4e (one O_2 molecule) $\times$ 2 light reactions) to reduce NADP and produce the necessary ATP at the same time.
  Nevertheless photosynthetic CO_2 fixation itself is only about 30% efficient (2.7 quanta/8–10 quanta) as we can measure it. Taken in conjunction with an average efficiency of less than 1% for whole plants capturing and utilizing photosynthetically active sunlight (see Chapter 1) this reinforces the concept that these energy exchanges can be very wasteful processes and could be improved.

6.5 Sucrose and starch synthesis

Sucrose is synthesized in the cytoplasm from triosephosphates (dihydroxyacetone phosphate and glyceraldehyde 3-phosphate) exported from the chloroplasts, via the chloroplast envelope, in exchange for orthophosphate (Pi). In the cytoplasm the triosephosphates combine to form fructose 1,6-bisphosphate, which is subsequently hydrolysed to fructose 6-phosphate by the enzyme fructose bisphosphatase. Fructose 6-phosphate is isomerized to glucose 1-phosphate via glucose 6-phosphate.

The synthesis of sucrose involves the participation of uridyl (U) phosphates and the following steps catalyzed by cytoplasmic enzymes.

$$\text{Glucose 6-phosphate} \longrightarrow \text{glucose 1-phosphate}$$

$$\text{Glucose 1-phosphate} + \text{UTP} \longrightarrow \text{UDP-glucose} + \text{pyrophosphate.}$$

There are two pathways for conversion of UDP-glucose to sucrose. In plants such as sugar cane:

$$\text{UDP-glucose} + \text{fructose 6-phosphate} \longrightarrow \text{sucrose 6-phosphate} + \text{UDP}$$

$$\text{Sucrose 6-phosphate} + H_2O \longrightarrow \text{sucrose} + \text{Pi}$$

In some other plants the following pathway exists:

$$\text{UDP-glucose} + \text{fructose} \longrightarrow \text{UDP} + \text{sucrose}$$

Under conditions where the CO_2 fixation rate in chloroplasts exceeds the rate at which triose phosphates can be converted to sucrose in the cytoplasm, synthesis of starch (a glucose polymer) occurs in the chloroplast stroma. The triose phosphates are converted to glucose 1-phosphate which then reacts with ATP as:

$$\text{Glucose 1-phosphate} + \text{ATP} \longrightarrow \text{ADP glucose} + \text{pyrophosphate.}$$

This reaction is catalysed by the enzyme ADP-glucose pyrophosphorylase.

$$\text{ADP-glucose} + [\text{glucose}]_n \xrightarrow{\text{starch synthase}} [\text{glucose}]_{n+1} + \text{ADP}$$

The starch synthesized in the chloroplasts is broken down, to sugars or sugar phosphates (depending on the availability of Pi), and utilized by the plant during periods of darkness or limited photosynthesis.

The partitioning of triose phosphates between sucrose synthesis in the cytoplasm and starch formation in the chloroplasts is regulated by a number of factors. The final step in the sucrose biosynthesis pathway in the cytoplasm liberates Pi and this Pi must be reimported to the chloroplast in order to export triose phosphates into the cytoplasm via the phosphate translocator in the chloroplast envelope (§8.10). Another potential control of sucrose synthesis is via the regulation of the activity of fructose bisphosphatase. This enzyme, which hydrolyses fructose 1,6-bisphosphate, is inhibited by fructose 2,6-bisphosphate, a cytoplasmic metabolite, whose concentration in leaves varies with the incident light intensity (§8.9). A third controlling factor is ADP glucose pyrophosphorylase, a key enzyme in starch biosynthesis, which is activated by phosphoglyceraldehyde (PGA) and inhibited by Pi; thus the rate of starch synthesis is dependent on the [PGA]/[Pi] ratio in the stroma.

Sucrose being a neutral (non-electrolyte), non-reducing sugar which is highly soluble in water serves as a useful metabolite regulating the osmotic and water balance between the various cellular constituents of the plant.

6.6   Light-coupled reactions of chloroplasts other than CO_2 fixation

In addition to CO_2 assimilation there are many other 'dark' reactions localized in the chloroplast stroma (and also envelope) which utilize ATP, $NADPH_2$, reduced Fd, and sugar phosphates generated by photosynthetic electron transport. Some of these light-coupled reactions which are essential for the synthesis of organic material by plants are outlined here.

(a) Assimilation of nitrogen   Nitrate is the most important nitrogen source for plants. The first step in the assimilation of nitrate is its reduction to nitrite catalysed by the enzyme nitrate reductase, the electrons being donated by reduced NAD or NADP.

$$NO_3^- + NAD(P)H_2 \longrightarrow NO_2^- + NAD(P) + H_2O$$

Nitrate reductase is present in roots as well as in leaves. Isolated chloroplasts do not reduce nitrate since the enzyme is present only in the cytosol. However, nitrate reduction in leaf extracts is stimulated in light due to an increased supply of reductants (possibly phosphoglyceraldehyde or malate) from the chloroplasts which generate $NADH_2$ in the cytosol ($NADPH_2$ produced in the chloroplasts cannot pass through the chloroplast envelope). Some species of blue-green algae contain a nitrate reductase which can take up electrons from reduced ferredoxin for the reduction of nitrate.

The enzyme nitrite reductase, which catalyses the reduction of nitrite to ammonia, is located in the stroma. Reduced ferredoxin is the electron donor for this reaction.

$$NO_2^- + 6e^- + 8H^+ \longrightarrow NH_4^+ + 2H_2O$$

Since ammonia is toxic to the plant it is immediately assimilated as glutamine via an ATP-dependent reaction catalysed by glutamine synthetase.

$$NH_3 + glutamate + ATP \longrightarrow glutamine + ADP + Pi$$

Many species of blue-green algae can fix atmospheric nitrogen to ammonia. This nitrogen assimilation is catalysed by the enzyme nitrogenase and utilizes electrons from reduced ferredoxin and energy from ATP – both products of photosynthetic electron transport.

(b) Assimilation of sulphate Reductive assimilation of sulphate occurs mainly in the leaves – sulphate probably enters the chloroplasts via a sulphate translocator in the chloroplast envelope (§8.10). In isolated chloroplasts sulphate assimilation is light-dependent and requires both reduced ferredoxin and ATP. Sulphate is initially reduced to sulphite (with the consumption of ATP) and further to sulphide via sulphite reductase, a ferredoxin-dependent enzyme localized in the chloroplast. Cysteine synthetase, another chloroplast enzyme, immediately assimilates sulphide as the amino acid cysteine which is then further metabolized to other biological sulphur compounds such as glutathione, methionine, Co enzyme A, sulpholipids, etc.

(c) Fatty acid biosynthesis Chloroplasts are the main site of fatty acid synthesis in leaves. Fatty acid biosynthesis is catalysed by the fatty acid synthases in the stroma and requires photosynthetically produced ATP and $NADPH_2$. The primary precursor is acetyl CoA (possibly derived from acetate) and the main products are the saturated fatty acids, palmitic (C_{16}) and stearic (C_{18}), and the unsaturated fatty acid, oleic ($C_{18:1}$).

(d) Oxygen exchange Oxygen is a by-product of photosynthetic water splitting leading to CO_2 fixation; the concentration of oxygen in the chloroplasts in the light is always higher than that of the air surrounding the leaf. Accumulation of oxygen will be toxic to the chloroplasts since it can cause oxidation of membrane lipids and inhibition of enzymes of the CO_2 fixation pathway such as ribulose bisphosphate carboxylase (by binding to the enzyme) and glyceraldehyde 3-phosphate dehydrogenase (by oxidation of –SH groups). Oxygen and its reduction products are metabolized in the chloroplasts by a number of pathways (Fig. 6.8). Photorespiration in C_3 plants (§6.7) is one such means of disposing of molecular oxygen. Chloroplasts contain a high (millimolar) concentration of reduced glutathione (GSH, glutamyl cysteinyl glycine) which can react with oxygen:

$$2GSH + \tfrac{1}{2}O_2 \longrightarrow 2GSSG + H_2O$$

The oxidized glutathione is reduced back to GSH by $NADPH_2$, a reaction catalysed by glutathione reductase, a stromal enzyme:

$$GSSG + NADPH_2 \longrightarrow 2GSH + NADP$$

Excited chlorophyll molecules in their triplet state (Fig. 4.2b) can transfer their energy to the oxygen molecule generating an excited oxygen species known as singlet oxygen. Singlet oxygen can damage thylakoid membrane structure by oxidizing polyunsaturated fatty acids to lipid peroxides. This oxidative damage

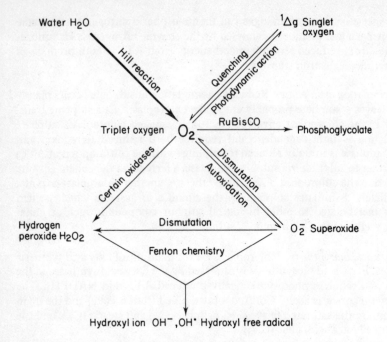

Fig. 6.8 Major pathways of oxygen metabolism in the active chloroplast. (From Foyer and Hall, 1980.)

is prevented by the removal of singlet oxygen by reductants in chloroplasts such as α-tocopherols, carotenoids, ascorbate, and GSH.

Monovalent reduction of oxygen by photoreduced PSI electron acceptors (reduced Fe–S centres and ferredoxin) has been demonstrated in intact chloroplasts and in leaves at rates varying between 3 to 27 percent of total electron transport, dependent of the irradiance and CO_2 concentration.

The primary product of such oxygen reduction by reduced ferredoxin is superoxide (O_2^-), a highly active free radical species:

$$O_2 + Fd_{red} \longrightarrow Fd_{ox} + O_2^-$$

The superoxide thus generated is converted to H_2O_2 via two different pathways:

$$O_2^- + 2H^+ + Fd_{red} \longrightarrow Fd_{ox} + H_2O_2 \tag{1}$$

$$2O_2^- + 2H^+ \longrightarrow O_2 + H_2O_2 \tag{2}$$

The second reaction, dismutation of superoxide, is catalysed by the Cu–Zn containing *superoxide dismutase* enzyme occurring in the chloroplast stroma.

Hydrogen peroxide is a toxic molecule and it is reduced to water by the catalytic action of metal ions and also by ascorbate present in high concentrations (up to 50 mM) in the chloroplasts:

$$2H^+ + H_2O_2 + \text{ascorbate} \xrightarrow[\text{peroxidase}]{\text{ascorbate}} 2H_2O + \text{dehydroascorbate}$$

As already mentioned (§5.4) the photoreduction of oxygen by reduced ferredoxin is coupled to phosphorylation of ADP – this type of noncyclic photophosphorylation is one of the major means of balancing the ATP/NADPH$_2$ ratio in the chloroplasts.

6.7 The C_4 pathway of CO_2 fixation

Many tropical grasses and plants such as sugar cane and maize are able to fix CO_2 initially into 4-carbon compounds like oxaloacetate, malate and aspartate in addition to CO_2 fixation by the Calvin C_3 cycle.

As already mentioned (§3.4) the leaves of these C_4 plants possess two types of chloroplasts, mesophyll and bundle sheath. The stomata of the C_4 plants are usually located such that the substomatal cavity is immediately adjacent to the mesophyll cell chloroplasts. The CO_2 which diffuses into the leaf through the stomata enters the mesophyll cytoplasm where it reacts with phosphoenol pyruvate [PEP] to form oxaloacetate in the presence of the enzyme PEP carboxylase.

$$CO_2 + \text{PEP [CH}_2 = \overset{\overset{\displaystyle OP}{|}}{C} - \text{COOH]} \xrightarrow[\text{carboxylase}]{\text{PEP}} \text{oxaloacetate}$$

There is a high concentration of PEP carboxylase in the mesophyll cells of C_4 species and thus CO_2 can be readily fixed down to low concentrations. The oxaloacetate is subsequently reduced by NADPH$_2$, formed by the normal light reactions, to malate. Radioactivity-labelling experiments using $^{14}CO_2$ have shown that more than 90% of the radioactivity is fixed in C_4 acids in one second. The malate is then transported to the bundle sheath cells where it is decarboxylated to pyruvate and CO_2 and the CO_2 thus released is then used for sugar and starch production via the Calvin C_3 cycle. The malate can also function as a constituent of the Kreb's cycle or can be aminated to aspartate and form a constituent of the amino acid pool. A simplified scheme of CO_2 fixation by C_4 plants is given in Fig. 6.9.

The rate of photosynthetic CO_2 fixation by C_4 plants is not affected by atmospheric concentrations of O_2 (which is high) and CO_2 (low) both factors which normally enhance the photorespiration rate of C_3 plants (see later). The water use efficiency, i.e. the ratio of the mass of CO_2 assimilated to water transpired, in C_4 plants is often twice that of C_3 species. Also salinity tolerance is a common feature of many C_4 species. All these traits allow C_4 plants to survive in

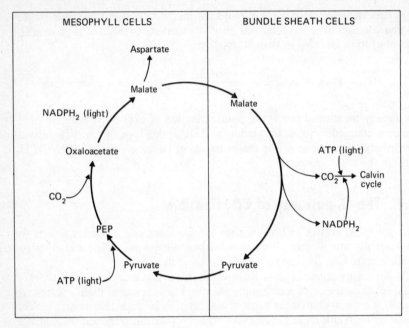

Fig. 6.9 CO_2 fixation scheme in C_4 plants.

dry and saline habitats. Though C_4 species are of tropical zone origin and of wide occurrence in drier zones their physiological adaptation to cool temperate zones has occurred. For example Woolhouse and Long have found many characteristics of C_4 plants in the marsh grass *Spartina townsendi* growing in the coastal areas of Britain.

6.8 Crassulacean acid metabolism

Many succulent plants growing in arid environments fix CO_2 in the dark to the C_4 acids oxaloacetic and then malic – the phenomenon was investigated extensively in the Crassulaceae and termed Crassulacean acid metabolism (CAM). CAM is widespread in the angiosperm families Agavaceae, Bromeliaceae, Cactaceae, Crassulaceae, Euphorbiaceae, Liliaceae, Orchidaceae, etc. CAM plants normally close their stomata during the day to prevent water loss. Their stomata open at night. CO_2 enters the leaves and combines with PEP (a product of starch metabolism) to form oxaloacetic acid in the presence of the enzyme PEP carboxylase which is found in the cytoplasm of the leaf cells. The oxalo-acetate is reduced by malic dehydrogenase to malic acid, which accumulates in the leaf vacuoles (Fig. 6.10). During the day the stomata become closed, the malate is transported to the cytoplasm where it is decarboxylated by a malic enzyme to yield pyruvate and CO_2. The CO_2 thus released enters the chloro-plasts where it is fixed to sugars by the photosynthetic Calvin C_3 cycle. Thus in

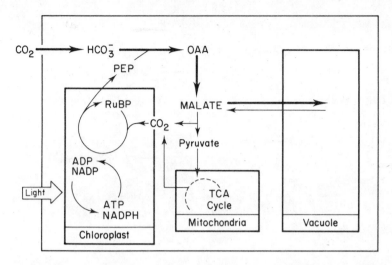

Fig. 6.10 Flow of carbon in CAM plants. Thick arrows indicate reactions which predominate in the dark, thin arrows indicate reactions which occur in the light. **OAA**, oxaloacetate; **PEP**, phosphoenol pyruvate; **RuBP**, ribulose bisphosphate; **TCA**, tricarboxylic acid. (From Coombs, Hall and Chartier, 1983.)

CAM plants the fixation of CO_2 to malate at night and its decarboxylation to CO_2 and pyruvate during the day are separated in time whereas in C_4 plants the two phases are separated spatially where the primary carboxylation occurs in the mesophyll and the decarboxylation reactions in the bundle sheath. Given adequate water CAM plants may behave like C_3 species.

6.9 Photorespiration and glycollate metabolism

An active field of current research is the study of photorespiration (the light-stimulated release of CO_2 at rapid rates by leaves) which is quite different from the 'dark' evolution of CO_2 by mitochondrial respiration in leaves. Plant species differ markedly in their rates of photorespiration; in some inefficient photosynthetic species it may be as high as 50% of net photosynthesis. Labelling experiments using the heavy oxygen isotope ^{18}O have shown that in plants with high photorespiration rates one of the initial products containing the ^{18}O label is the two carbon (C_2) acid, glycollic acid, followed in time by glycine, serine and 3-phosphoglyceric acid [PGA], a Calvin cycle intermediate. The proposed metabolic pathway of glycollic acid oxidation which leads to CO_2 formation (during the conversion of glycine to serine) is shown in Fig. 6.11. There may be additional oxidation reactions producing CO_2. How is glycollic acid formed in the chloroplasts? The enzyme ribulose bisphosphate carboxylase (RuBisCO) which is the major protein of the chloroplast stroma has a molecular weight of 500–560 kD and contains eight copies of two types of polypeptides designated as the large subunit (LSU) and the small subunit

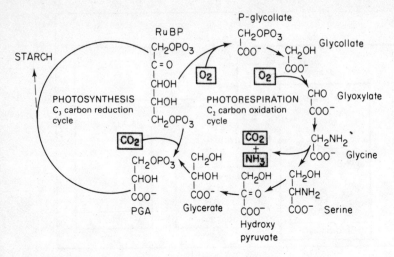

Fig. 6.11   Integrated carbon reduction and photorespiratory carbon oxidation cycle. (After Lorimer, G.H. *et al.*, see Rao and Hall, 1982.)

(SSU). The catalytic activity of the enzyme is associated with the LSU (Mr 50 to 55 kD) which is coded by the chloroplast DNA and synthesized in the chloroplasts. The SSU (Mr 12 to 15 kD) whose function is not well understood is coded by the nuclear DNA, synthesized in the cytoplasm as a precursor, and transported into the chloroplasts (§8.2). When the enzyme functions as a carboxylase of RuBP in the Calvin cycle CO_2 acts as a substrate as well as an activator of the enzyme. In addition to its carboxylase action RuBisCO also functions as an oxygenase and catalyses the oxygenation of RuBP to 2-phosphoglycollic acid and PGA. The phosphoglycollic acid is then hydrolysed to phosphate and glycollic acid.

Table 6.1 General characteristics of C_3, C_4 and CAM plants (from Edwards and Walker, 1983).

	C_3	C_4	CAM
1.	Typically temperate species e.g. spinach, wheat, potato, tobacco, sugar beet, soya bean, sunflower.	Typically tropical or semi-tropical species e.g. maize, sugar cane, Amaranthus, Sorghum, savannah grasses. Plants adapted to high light, high temperatures and also semi-arid environments.	Typically arid zone species e.g. cacti, orchids, Agave, succulent plants.
2.	Moderately productive. Yields of 30 t (tonnes) dry weight per hectare (2.47 acres) possible. (Sunflower is highly productive.)	Highly productive, 80 t per hectare for sugar cane is possible.	Usually very poorly productive. (Pineapple is highly productive.)
3.	Cells containing chloroplasts do not show Kranz-type anatomy and generally lack peripheral reticulum. Only one type of chloroplast.	Kranz-type anatomy and peripheral reticulum are essential features. Often have two distinct types of chloroplasts.	Lack Kranz anatomy and peripheral reticulum. Only one type of chloroplast.
4.	Initial CO_2 acceptor is ribulose bisphosphate (RuBP), a 5 carbon sugar.	Initial CO_2 acceptor is phosphoenol pyruvate (PEP), a 3 carbon acid.	CO_2 acceptor is PEP in the dark and RuBP in the light.
5.	Initial CO_2 fixation product is the 3-carbon acid phosphoglycerate.	Initial CO_2 fixation product is the 4-carbon acid oxaloacetate.	CO_2 fixation products are oxaloacetate in the dark and phosphoglycerate in light.
6.	Only one CO_2 fixation pathway.	Two CO_2 fixation pathways separated in space.	Two CO_2 fixation pathways separated in time.
7.	High rates of glycollate synthesis and photo-respiration.	Low rates of glycollate synthesis; no photorespiration.	Same as C_4.
8.	Low water use efficiency and salinity (ion) tolerance.	High water use efficiency and salinity tolerance.	Same as C_4.
9.	Photosynthesis saturates at 1/5 full sunlight.	Do not readily photosaturate at high light.	Same as C_4.
10.	High CO_2 compensation point.	Low CO_2 compensation point.	High affinity for CO_2 by night.
11.	Open stomata by day.	Open stomata by day.	Open stomata by night.

Both CO_2 and O_2 compete for RuBP at the same site of the enzyme so that high concentrations of CO_2 and low concentrations of O_2 favour carboxylation, whereas high concentrations of O_2 and low concentrations of CO_2 (as is found in the atmosphere) favour oxygenation and thus formation of phosphoglycollic acid. It has also been reported that higher temperatures favour the oxygenation reaction by the enzyme. Glycollic acid can be formed by other metabolic pathways also in the chloroplast.

The external symptoms of photorespiration are (1) the inhibition of photosynthesis by increased oxygen, (2) the existence of a high CO_2 compensation point (30–50 ppm CO_2 at 25°C in air), and (3) the variation of the CO_2 compensation point in response to variations in O_2, light and temperature.

Plants belonging to the C_4 and CAM species do not show these symptoms though the kinetic properties of RuBisCO isolated from these species are similar to those of this enzyme from C_3 plants or even from photosynthetic bacteria. In the C_4 plants most of the RuBisCO is compartmentalized in the chloroplasts of bundle sheath cells which are not themselves in direct equilibrium with the atmospheric CO_2 or O_2. The CO_2 concentration in the bundle sheath cells may be much higher than that in the atmosphere since the CO_2 is produced *in situ* by the decarboxylation of malate which is imported from the mesophyll cells (see §6.7) – this allows carboxylation to compete more effectively with oxygenation for the enzyme-bound RuBP. Also, any CO_2 generated by photorespiration in the C_4 species could be trapped in the chloroplasts as a result of internal recycling by the PEP carboxylase of the mesophyll cells so that the loss of CO_2 to the atmosphere is prevented. The CAM plants synthesize a high level of malate at night. Since decarboxylation of malate and CO_2 fixation by the C_3 pathway in CAM species occurs in the day when the stomata are closed (see §6.8), the internal concentration of CO_2 in these plants may well be much higher than that of the external atmosphere thus stimulating carboxylation over oxygenation. It has been shown that if C_3 plants are placed in artificial atmospheres of high CO_2 and low O_2 pressures they will behave like C_4 plants in having low photorespiratory activity.

7
Bacterial Photosynthesis

7.1 Classification

Photosynthetic bacteria are typically aquatic micro-organisms inhabiting marine and freshwater environments like moist and muddy soil, stagnant ponds and lakes, sulphur springs, etc. There are three major types.

1. *Green bacteria*. These are subdivided into two families, viz. the Chlorobiaceae and the Chloroflexaceae. The Chlorobiaceae are strict anaerobes which grow by utilizing sulphide or thiosulphate as electron source, e.g. *Chlorobium limicola* and *Ch. thiosulfatophilum*. The Chloroflexaceae are facultative aerobes which can utilize reduced carbon compounds as electron donors for their growth, e.g. *Chloroflexus aurantiacus*. The two families are similar in their photosynthetic pigment composition and membrane fine structure but have different types of electron transfer pathway.
2. *Purple sulphur bacteria* (Chromatiaceae) which can use hydrogen sulphide as a photosynthetic electron donor, e.g. *Chromatium*.
3. *Purple non-sulphur bacteria* (Rhodospirillaceae) which are unable to use hydrogen sulphide and depend on the availability of simple organic compounds like alcohols and acids as electron donors, e.g. *Rhodomicrobium, Rhodobacter, Rhodopseudomonas* and *Rhodospirillum*.

When grown photosynthetically all three types are strict anaerobes, i.e. grow only in the complete absence of oxygen. They cannot use water as a substrate and they do not evolve oxygen during photosynthesis.

7.2 Photosynthetic pigments and apparatus

The pigment systems of photosynthetic bacteria are slightly different from those of plants and algae. The chlorophyllous pigments of bacteria are called *bacteriochlorophylls*; five classes of bacteriochlorophylls (B Chl *a*, B Chl *b*, B Chl *c*, B Chl *d* and B Chl *e*) have been characterized. These bacteriochlorophylls are very similar to chlorophylls *a* and *b* but differ in the nature of the side chains attached to the carbon atoms 2, 3, 4, 5, 7 and 10 shown for chlorophyll in Fig. 3.5. In addition a magnesium-less bacteriochlorophyll

Table 7.1 Characteristics of photosynthetic bacteria.

Group	Photosynthetic pigments	e^- donor (substrate for growth)	Growth conditions and other properties	Examples
Greens (a) Chlorobiaceae	B Chl a plus B Chl c, B Chl d or B Chl e. Carotenoids Reaction centre P840	H_2S $Na_2S_2O_3$ H_2	Light, autotrophic; strict anaerobes; non motile PS apparatus: Chlorobium vesicles and associated membranes	*Chlorobium limicola*, *Ch. thiosulfatophilum*, *Prosthecochloris aestuarii*
(b) Chloroflexaceae	Similar to (a)	Organic substrates	Facultative aerobic; filamentous gliding	*Chloroflexus aurantiacus*
Purple sulphur bacteria (Chromataceae)	B Chl a or B Chl b Carotenoids Reaction centre P870 or P890	H_2S, $Na_2S_2O_3$ H_2 Organic substrates e.g. acetate	Autotrophic and heterotrophic in light. Strict anaerobes. PS apparatus: chromatophores	*Chromatium D* *Thiocapsa roseopersicina*
Purple non-sulphur bacteria (Rhodospirillaceae)	B Chl a or B Chl b Carotenoids Reaction Centre P870 or P960	Organic substrates e.g. succinate, malate H_2	Heterotrophic or autotrophic in light and anaerobic. Will grow aerobically and heterotrophically in the dark	*Rhodospirillum rubrum* *Rhodopseudomonas viridis*, *Rhodobacter spheroides*

called bacteriopheophytin is found in the reaction centre of all photosynthetic bacteria. The principal carotenoids of photosynthetic bacteria are also slightly different chemically from the algal carotenoids. The nature of some of the pigments found in the photosynthetic bacteria and their growth requirements are given in Table 7.1. The absorption spectra of two typical bacteria are shown in Fig. 7.1.

The photosynthetic apparatus in purple and green bacteria are of morphologically different types and both types are distinct from the photosynthetic unit found in chloroplasts. The action spectra of bacteriochlorophyll fluorescence in purple bacteria indicate that light energy absorbed by carotenoids and shortwave bacteriochlorophyll bands is transferred to the longest wavelength bacteriochlorophyll (which absorb at 870, 890 and 960 nm) before being used for photosynthesis. From measurements of substrate (carbon) assimilation and photosynthetic phosphorylation by suspensions of purple bacteria, during flashing light experiments, Clayton has estimated that the bacterial photosynthetic unit contains 30 to 50 bacteriochlorophyll molecules; these may vary from species to species.

When cells of purple bacteria are disrupted they release a class of subcellular particles containing all the photosynthetic pigments. These pigment-bearing particles can be isolated by differential centrifugation. When examined by electron microscopy after staining the particles appear like spherical bodies 30 to 100 nm in diameter and are called *chromatophores*. Each chromatophore contains several photosynthetic units. The chromatophores prepared from *Rhodobacter* (formerly *Rhodopseudomonas*) *sphaeroides*, for example, consist of approximately 40 reaction centre (RC) complexes, 500 light harvesting (LH) complexes, 1000 carotenoids and 1000 ubiquinone molecules. They are probably derived from the external cytoplasmic membrane by extensive invaginations (infolding) of the membrane.

Duysens showed that the absorption spectrum of bacteriochlorophyll in

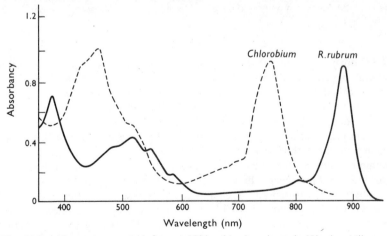

Fig. 7.1 Absorption spectra of green (*Chlorobium*) and purple (*Rhodospirillum rubrum*) photosynthetic bacteria.

purple bacteria is changed reversibly by illumination. The change corresponds mainly to a bleaching (oxidation) of the long wave absorption band of bacterio-chlorophyll; at 890 nm in *R. rubrum* and *Chromatium*, and at 870 nm in *Rb. sphaeroides* – these are now called reaction centre Bchls.

By treating chromatophores of *Rb. sphaeroides* with detergents and then illuminating them in oxygen it is possible to destroy the light-harvesting bacteriochlorophyll molecules while still keeping the light-reacting component intact. The 870 nm absorption band of such specially prepared chroma-tophores is oxidized reversibly by light. This special light-reacting component in *Rb. sphaeroides* is called P870 and its role is similar to that of P700 in chloro-plasts. The equivalent component in *R. rubrum* and in *Chromatium* is desig-nated P890. There is one P870 or P890 molecule for about 40 bacterio-chlorophyll molecules which constitute the photosynthetic unit (Fig. 7.2). Difference spectroscopic studies (see Chapter 4) showed the reversible oxidation-reduction of a quinone (ubiquinone) and of a cytochrome simultaneously with the light-induced changes in P870.

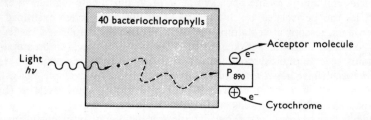

Fig. 7.2   Diagrammatic representation of the photosynthetic unit in bacteria. One photon of light reacts in a unit of 40 bacteriochlorophyll molecules containing one P890 reaction centre.

The photosynthetic pigments of the green bacteria are located predominantly in special structures called the *Chlorobium vesicles* or *chlorosomes* which are firmly attached to the cytoplasmic membrane. The chlorosome pigments constitute the bulk of the very large antenna found in all green bacteria, which may be as high as 1500 bacteriochlorophylls per reaction centre.

In recent years due to improvements in the techniques of isolation of bacterial reaction centres and the introduction of picosecond (10^{-12} s) laser flash spectro-photometry, most of the events occurring in the light reactions of bacterial photosynthesis have been elucidated. Light energy absorbed by bacterio-chlorophylls and carotenoids is channelled to a reaction centre containing a few (2 or 4) specialized BChl molecules. Charge separation occurs across the membrane at these BChl molecules, followed by electron trans-port resulting in the production of ATP, $NADH_2$ or reduced ferredoxin (Fig. 7.3).

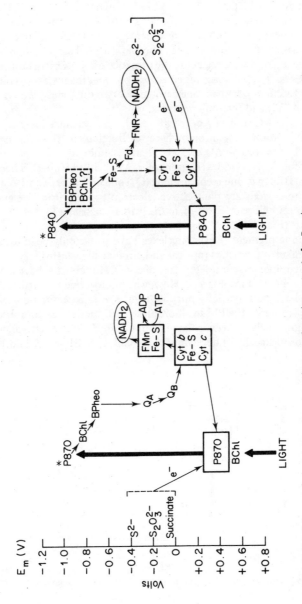

Fig. 7.3 Scheme of electron transport in photosynthetic bacteria. **Left:** Purple; **Right:** Green.

7.3   Photochemistry and electron transport

By selective use of different types of detergents it is possible to isolate pure reaction centres from chromatophores and study their spectral and structural properties. The most fully characterized reaction centre is that from *Rb. sphaeroides*. It has an Mr of approximately 80 kD and is built up of three poly-peptides of Mr 21 kD, 24 kD and 32 kD designated L, M and H subunits respectively. Each reaction centre contains 4 BChl, 2 BPheo, one ferrous iron, two ubiquinones (one loosely and the other tightly bound), and one carotenoid. The composition of reaction centres of other purple bacteria are more or less similar – the major difference being the presence of *c*-type cytochromes in some RC complexes. The RC unit of *Thiocapsa pfennigi* for instance, contains one P960, four BChl, two BPheo, one carotenoid, two quinones, two cytochromes (*c*-555 and *c*-557) and five polypeptides.

Recently, the RC unit of *Rps. viridis* has been crystallized and its structure determined at atomic resolution by X-ray diffraction analysis. The crystal unit (3 × 7 × 13 nm) has an Mr of 150 kD and is embedded in a protein moiety made up of four polypeptides (Deisenhofer, Michel and Huber). There are four BChl *b*, two BPheo *b*, one menaquinone (Q_A) one ubiquinone and a *c*-type cyto-chrome. The crystals are photoactive, electron transfer from the energy trap bacteriochlorophyll pair $(BChl-b)_2$ to the primary acceptor Q_A occurring in 250 pico seconds (Fig. 7.4).

The antenna pigment (LH) complexes have so far only been identified by their spectral characteristics (optical and circular dichroism).

Pure RC complexes, completely free of core LH pigments, have not yet been prepared from the Chlorobiaceae. However, photoactive RC structures bound to proteins have been partially purified from *Prosthecochloris aestuarii* and are found to be similar to the PSI in chloroplast. Fully active RC complexes free of core LH pigments have been isolated from *Chloroflexus aurantiacus*, a member of the Chloroflexaceae. They contain 3 BChl *a*, 3 BPheo *a*, and two poly-

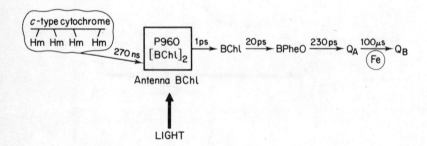

Fig. 7.4   Light-induced electron transfer sequence and time scales ($t\frac{1}{2}$) in the reaction centre of *Rhodopseudomonas viridis*. **BChl**, bacteriochlorophyll; **BPheo**, bacteriopheophytin; **Hm**, haem; **Q**, quinone; (After Deisenhofer, Michel and Huber, 1985.)

peptides per RC and are structurally similar to the RC complexes isolated from purple bacteria. The photosynthetic electron transport pathway of *Chloroflexus* also is similar to that of purple bacteria.

7.4 Carbon dioxide fixation

Photosynthetic bacteria do not show an Emerson enhancement effect, so it is generally considered that there is only one major photoreaction in photosynthetic bacteria. In cell-free preparations of photosynthetic purple bacteria the dominant photosynthetic reaction is cyclic photophosphorylation (production of ATP). Particles have been prepared from green bacteria which are able to catalyse photoreduction of ferredoxin (see Chapter 5) which can subsequently be used to reduce NAD to $NADH_2$ via a flavoprotein enzyme. Thus the photosynthetic bacteria can generate both ATP (energy) and $NADH_2$ (reducing power) for the fixation of CO_2 (Fig. 7.3). The Calvin–Benson cycle (Chapter 6) enzymes are shown to be present in nearly all strains of photosynthetic bacteria studied which thus have the ability to fix CO_2 via this cycle. Evans, Buchanan and Arnon have demonstrated the existence of a different route for photosynthetic CO_2 fixation in bacteria which is mediated by reduced ferredoxin leading to the synthesis of α-keto acids, e.g. pyruvate and α-ketoglutarate. Their scheme of reductive carboxylic acid cycle in photosynthetic bacteria is shown in Fig. 7.5. In this pathway of photosynthetic CO_2 fixation

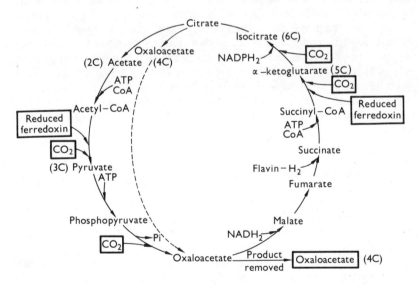

Fig. 7.5 The reductive carboxylic acid cycle of photosynthetic bacteria. Four CO_2 molecules are fixed to form one oxaloacetate molecule. (Scheme devised by Evans, Buchanan and Arnon (1966). *Proceedings of the National Academy of Science, U.S.,* **55**, 928.)

the ferredoxin (Fd)-dependent enzymes pyruvate synthase and α-ketoglutarate synthase catalyse the carboxylation of acetyl and succinyl-coenzyme A, respectively, as shown:

1. Acetyl CoA + CO_2 + Fd_{red} $\xrightarrow{\text{enzyme}}$ Pyruvate + CoA + Fd_{ox}

2. Succinyl CoA + CO_2 + Fd_{red} $\xrightarrow{\text{enzyme}}$ α-ketoglutaric acid + CoA + Fd_{ox}

The overall cycle involves the fixation of four molecules of CO_2. It is probable that in these bacteria both the reductive pentose phosphate (Calvin) pathway and the reductive carboxylic acid cycle are operating in photosynthetic CO_2 fixation.

7.5 Ecological significance of phototrophic bacteria

Under anaerobic (O_2–free) conditions organic matter is fermented by various micro-organisms (the chemosynthetic anaerobes) which gain their energy by a substrate-linked phosphorylation. Various metabolic end products like CO_2, H_2, ethanol and simple fatty acids are formed during this process. Such compounds would accumulate if they were not removed as nutrients by other types of microbes which are unable to use oxygen as the ultimate electron acceptor in their respiratory processes. The sulphate- and nitrate-reducing bacteria are able to consume part of the end products of fermentation of the chemosynthetic anaerobes. The phototrophic bacteria (green and purple photosynthetic bacteria) derive their energy from light and are able to meta-bolize most of the end products of anaerobic fermentation like alcohols, acids and hydrogen, as well as the end products of sulphate and nitrate respiration such as H_2S and N_2. Thus the cell materials synthesized by the green and purple photosynthetic bacteria are future substrates for the chemosynthetic anaerobes which again produce the nutrients for the phototrophic bacteria. Thus these two types of bacteria growing in an O_2-free environment can exist together.

Light energy conversion by halobacteria

A unique type of light energy conversion for ATP synthesis is shown by certain halobacteria typified by *Halobacterium halobium*. These organisms normally grow in aerobic, extremely saline environments (3.5 to 5M NaCl) – however, when cultured at low oxygen concentrations and high light intensity *H. halobium* cells form patches of a purple membrane on their surface which contain a single protein, bacteriorhodopsin. Bacteriorhodopsin can use the visible spectrum of light energy to generate a proton gradient from the interior of the cell to the exterior. This transmembrane proton gradient is coupled via an ATPase for the synthesis of ATP by the bacterium. Halobacteria lack antenna pigments and the carotenoids of *H. halobium* cannot transfer light energy to bacteriorhodopsin. So, the efficiency of energy conversion is probably lower than that in chlorophyll-mediated photosynthesis.

7.6 A comparison of plant and bacterial electron transport: evolution of photosynthesis

A close examination of the electron transport chains in chloroplasts (Fig. 5.2) and in the two bacterial photosystems (Fig. 7.3) reveals many common features shared by all three photosystems. Photosynthetic bacteria, algae, and plants regulate the amount of light-harvesting components synthesized in response to environmental light intensity conditions. All LH (light harvesting antenna pigments) are conjugated to proteins. The main distinction among light-harvesting properties of photosynthetic cells lies in the patterns of pigment localization. In the green-sulphur bacteria the light-harvesting function is located in non-membranous particulate structures, chlorosomes, embedded in the cell membrane and whose numbers and structures vary with the light intensity. In the purple bacteria the LH and RC complexes are together within a membrane system. In cyanobacteria and red algae the water-soluble phyco-bilins are found in the phycobilisomes localized on the outer surface of the photosynthetic membranes. The phycobilisomes contain about 300 to 800 chromophores which absorb light over much of the visible spectrum. The energy absorbed by the phycobilisomes is transferred to the RC with an efficiency approaching 100%.

The RC of bacteria and plants are membrane-bound proteins containing Chl or BChl. The primary photoreaction in all systems is a light-induced electron transfer from a donor species D to an acceptor A. The initial donor is a specialized Chl a molecule in plants and a BChl dimer in bacteria; the primary acceptor appears to be a Pheo or BPheo: Fe–quinone complex. The relatively low energies of excited Chl a and BChl at the RC helps to minimize the wasteful decay of the excited states via fluorescence with the result that the quantum yields of electron transfer reactions are high at the RC complex. Wasteful back reactions are also prevented by conducting the electron transfer from D* through a series of donor-acceptor species stabilized for increasingly longer lifetimes and separated by increasingly longer distances.

There are highly conserved patterns of sequences of amino acids which are common to the L and M subunits of the RC proteins of *Rps. capsulata* and the 32 kD herbicide binding protein of PSII isolated from spinach and tobacco leaves. The homology between these polypeptides suggest a common ancestor for these bacterial and plant proteins.

In plant as well as bacterial photosystems a quinone-cyt b–cyt c (cyt f) assembly functions as a gate between a two electron carrier and one electron carriers via a Q-cycle in which protons are transported across the membrane. The final product of photosynthetic electron transport in the green-sulphur bacteria is reduced pyridine nucleotide (NADH) as also is NADPH in the PSI of plants. In this respect, and in the content and organization of electron transfer intermediates the green bacterial photosystem (*chlorobiaceae*) resembles plant PSI. The purple bacterial photosynthesis, on the other hand, is more analogous to PSII of plants.

The biochemical evolution of energy-transducing systems is generally considered to have followed the sequence:

$$\text{fermentation} \longrightarrow \text{photosynthesis} \longrightarrow \text{respiration}.$$

Anaerobic fermentations probably were the earliest energy supplying processes and some of the catalysts and products of fermentations such as ferredoxins, ATP and NADH are constituents of the energy metabolism of every other type of organism. Gest has proposed an hypothesis to explain the mechanism of transition from heterotrophy to phototrophy based on studies of the redox balances of fermentations in photosynthetic and non-photosynthetic bacteria (see Rao *et al.*, 1985). He uses the term 'electrophosphorylation' to designate any process in which electron flow in membranes is the driving force for phosphorylation of ADP. The establishment of a membrane-associated electron flow resulted in energy conservation through an electrophosphorylation module with Fe–S proteins, quinones and cyt *b* participating as catalysts. According to Gest the earliest photophosphorylation system could have evolved by the fusion of a membrane-bound pigment (Mg-porphyrin) complex into a fermentative anaerobe operating an electrophosphorylation module. The development of a 'photosynthetic unit' could result from a photoactivatable pigment complex being embedded vectorially in the membrane. Thus organic molecules could be replaced as an electron source for energy conversion precipitating a significant event in the course of evolution – one that linked biological systems to the inexhaustible energy supply from the sun.

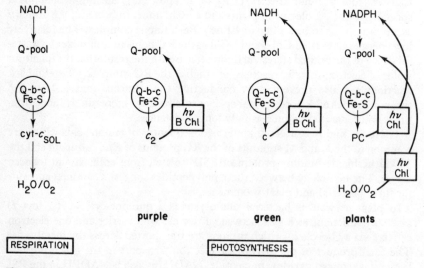

Fig. 7.6 A comparison of the respiratory and photosynthetic electron transport chains; see also Figs 5.2a and 7.3 (From Rao, Cammack and Hall, 1985.)

Once the photosynthetic pigment was incorporated into the cell membrane the next step would have been to devise a control mechanism to coordinate the energy flow from the extremely fast photochemical reactions to the rather slower biochemical reactions of the cell. The photosynthesizers solved this problem, as described earlier, by stabilizing the light-induced separation of charges via a cyclic electron flow through a series of intermediates of increasing life time. The addition of a *c*-type cytochrome to the electron flow cycle completed the evolution of the prototype of the purple sulphur bacteria. The unity of the bioenergetic processes is evident when one compares the respiratory and photosynthetic electron transport chains (Fig. 7.6). The Q-*b*-*c* complex plays a pivotal role in all electron transport chains, probably in conjunction with an Fe–S protein. Electrons are fed into the complex via a quinone pool and leave via a *c*-type cytochrome in respiratory and photosynthetic bacteria, and via plastocyanin (or a cytochrome in a few instances) in cyanobacteria and plants.

Another possibility is that the phototrophs evolved before the fermentative heterotrophs; thus all aerobic respiratory electron transport chains originated through modifications of the electron transfer chain in photosynthetic bacteria.

8

Research in Photosynthesis

The main areas of current research into the broad field of photosynthesis are summarized below. Many of these are interrelated and share research methodologies.

8.1 Protoplasts and cells

Protoplasts are basically cells without their cell walls (which have been removed by enzymatic digestion). Interferring substances, such as phenolics, present in cell walls or vascular tissues of leaves are removed during enzymatic digestion and hence chloroplasts prepared from protoplasts show a high degree of photosynthetic activity compared to the chloroplasts isolated by mechanical disruption of leaves. The refinement of protoplast isolation techniques has extended the range of plants from which active chloroplasts may be isolated. Thus, intact chloroplasts can now be isolated from C_3 grasses (wheat and barley), C_3 dicotyledons (tobacco, sunflower) as well as many C_4 species (maize, sugarcane). In C_4 species controlled enzymic digestion of leaves results in the preferential release of mesophyll protoplasts, leaving the bundle sheath strands intact. Protoplasts have proved very valuable in the study of inter- and intracellular compartmentation of photosynthetic metabolite activities in C_3, C_4 and CAM species of plants. Individual cells (with intact cell walls) can now also be isolated from the leaf in order to study the more complex interactions in the movement of substrates, ions, gases, etc. The techniques used vary greatly depending on the species and variety of plant being studied.

8.2 Origin and development of chloroplasts; chloroplast genetics

Did the chloroplasts of algae and higher plants (eukaryotes) evolve from the prokaryotic blue-green algae (cyanobacteria) as proposed in the endosymbiotic theory of evolution of eukaryotes? Are the *cyanelles* (symbiotic algae, see Fig. 3.4, p. 22) which are found in some obligate photoautotrophs such as the *Cyanophora* sp. (a flagellated algae containing a blue-green algal symbiont) modern representatives of evolutionary links between blue-green algae and chloroplasts? There is increasing evidence supporting the close similarity between blue-green algae and the chloroplasts of eukaryotes.

Chloroplast development

In angiosperms (flowering plants) chloroplast biogenesis is usually studied by germinating seeds in total darkness and then exposing the seedlings to light. The changes occurring during illumination are followed by examination of the ultrastructure of the plastids by electron microscopy, the assay of photosynthetic electron transport and CO_2 fixation rates, and determination of the chromophore-containing components, e.g. cytochromes and ferredoxins, by spectroscopy. Such studies have shown that chloroplasts develop from relatively undifferentiated structures known as proplastids via simple colourless organelles termed etioplasts.

The most important change occurring during illumination (and greening) is the reduction of protochlorophyllide to chlorophyllide (chlorophyll without the phytyl side chain). The response of leaves to light involves changes in the expression of specific genes at the RNA level – RNAs complementary to many segments of the plastid genome (see below) are expressed more abundantly in light. Substantial increases in the content of carotenoids, cytochromes, ferredoxin, plastocyanin, ferredoxin-NADP reductase, etc. have been measured during the greening process. Photosystem I activity is expressed at an earlier stage than PSII activity. The enzymes of the CO_2 fixation pathway are present in the etioplasts, albeit at a very low amount compared to mature chloroplasts. In higher plants the responses to light are thought to be mediated by three photoreceptors: protochlorophyllide, phytochrome (a coloured plant protein), and a blue light receptor.

Chloroplast genetics

The presence of DNA and ribosomes in chloroplasts were first demonstrated in 1962. Chloroplast DNA (plastome) occurs as a double stranded circular molecule with a unique base sequence. The average molecular mass of the plastome is about 1×10^8 kD. The molecule has a potential coding capacity for about 125 proteins. Features which distinguish the plastome from the nucleosome (nuclear DNA) are the presence in the former of a large segment of inverted repeating sequence and the absence of histones complexed to the DNA. The stromal and thylakoid ribosomes resemble prokaryotic ribosomes in many of their properties. Proteins coded by the plastome are synthesized on the chloroplast ribosomes. The identity of about 30 polypeptides coded by the plastome are known. These include the large subunit (LSU) of RuBisCO, the Q_B (herbicide-binding) protein, the cytochromes (f, b_6 and b_{559}), subunits of the ATP synthase complex (α, β and γ of CF_1 and two subunits of CF_0), some polypeptides associated with PSI and PSII, and many ribosomal proteins.

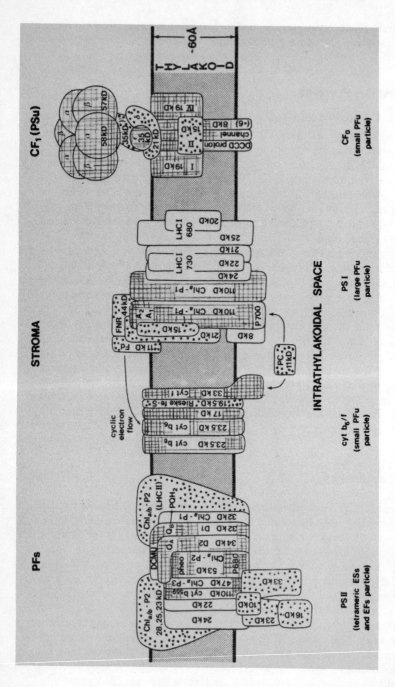

Fig. 8.1 Model of the photosynthetic membrane in barley thylakoids. Dotted proteins are coded by nuclear genes, synthesized on the cytoplasmic ribosomes and transported across the chloroplast envelope into the photosynthetic membrane. Crossed proteins are encoded by genes of the chloroplast DNA and synthesized on chloroplast ribosomes. The site of synthesis and gene localization for other proteins (with no dots and crosses) are still not known. See also Fig. 5.2. (Courtesy of Dr. D. Simpson and Prof. D. von Wettstein, Carlsberg Laboratory, Copenhagen.)

Much progress has been made in the last few years in identifying the loci of genes coding various proteins and various RNAs in the plastome. The techniques used are the digestion of DNA with site-specific restriction endonucleases, separation of the fragments by gel electrophoresis, cloning the purified fragments and in some cases determining their nucleotide sequences. The genetic nature of a fragment or cloned segment can be examined by its ability for *in vitro* transcription and translation of a specific chloroplast protein (which may be identified by immunoprecipitation with the protein antibody or by two-dimensional gel electrophoresis) or a specific RNA (identifiable by hybridization techniques or gel electrophoresis profiles). Such studies have resulted in the identification of the genes for two chloroplast proteins, Q_B and LSU in the plastome. The cloned gene of barley chloroplast LSU has been used to advantage in determining the amino acid sequence of the protein subunit and also in determining the binding sites of the substrate CO_2 molecule and the effector CO_2 molecule on the protein – both are bound to the lysine residues in the sequence. The genetic maps of chloroplast DNA from spinach and tobacco are shown in Fig. 8.2. These studies are in their early stages but progress is rapid and will undoubtedly lead to practical applications in crop improvements and genetic engineering in plants.

Transport and assembly of cytoplasmically-assembled polypeptides into the chloroplast membranes

The majority of the protein components of chloroplasts are encoded in the nuclear DNA and are synthesized on cytoplasmic ribosomes outside the organelle and subsequently imported into the chloroplasts. *In vitro* studies using the protein-synthesizing machinery of a wheat germ extract coupled to polyadenyl messenger RNAs (PolyA$^+$mRNA) isolated from algal or higher plants have demonstrated the transfer of a number of nuclearly-coded polypeptides into the chloroplast; examples are ferredoxin, plastocyanin, ferredoxin-NADP reductase, polypeptides associated with antenna pigment complexes, the small subunit of ribulose-bisphosphate carboxylase, and fructose 1-6-biphosphatase. These proteins are assembled on the cytoplasmic ribosomes as a precursor of much higher molecular weight, referred to as the transit peptide, with sequence extension, at the aminoterminal. The precursor of plastocyanin (Mr = 10.5 kD), for example, is assembled with a transit peptide of molecular mass 15 kD. The precursors are bound to the outer chloroplast membrane, translocated through the chloroplast envelope, processed to the mature polypeptide (possibly by proteolytic cleavage of the transit peptide), and then either assembled as membrane constituents or incorporated into the stroma. The import of precursors into the chloroplast is an energy-dependent process and consumes ATP synthesized by photophosphorylation. The requirements for binding of precursors to the envelope membrane, the nature of the receptors, and the mechanisms regulating the translocation and assembly of polypeptides are now being actively investigated.

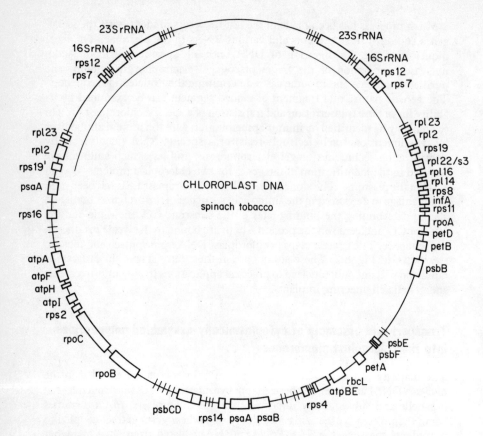

Fig. 8.2   Map showing the loci of genes coding for various proteins and RNAs identified in the circular chloroplast of DNA of spinach (**left**) and tobacco (**right**). Atp, ATP synthase; **inf**, initiation factor; **pet**, photosynthetic electron transfer components; **psa**, photosystem I proteins; **psb**, photosystem II proteins; **rbc**, large subunit of RuBisCO; **rpo**, RNA polymerase; **rps/rpl**, ribosomal protein. (Courtesy of Dr. J. C. Gray, University of Cambridge.)

Thus chloroplast biogenesis and function requires the cooperation of nuclear and chloroplast genomes. The biochemical control mechanisms which regulate the interaction between the chloroplast and nucleus are still poorly understood.

8.3   Chloroplast structure

Significant advances have been made towards the understanding of the molecular structure of the thylakoid membrane, i.e. the proteins and lipids and their orientation, the polypeptides associated with light-harvesting pigment–protein complexes and their distribution in the core and peripheral parts of the antennae, etc. Membrane complexes can be isolated from higher

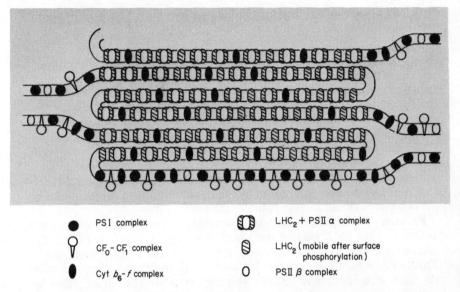

●	PS I complex	⬭⬭	LHC$_2$ + PSII α complex
⬯	CF$_0$ - CF$_1$ complex	⬭	LHC$_2$ (mobile after surface phosphorylation)
●	Cyt b_6- f complex	○	PSII β complex

Fig. 8.3 Schematic arrangement of light harvesting and electron transfer complexes in the thylakoid membrane. (Courtesy of Professor J. Barber, Imperial College, London.)

plant chloroplasts and their photochemical properties studied *in vitro*. There is increasing evidence of a lateral heterogeneity in the distribution of these complexes between stacked and unstacked regions of the chloroplast membrane (Fig. 8.3). Photosystem II itself is found to exist as two structurally and functionally independent types designated as PSII$_\alpha$ and PSII$_\beta$ – PSII$_\alpha$ located in the grana membranes and PSII$_\beta$ exclusively located in the stroma-exposed thylakoids. The light harvesting capacity of PSII$_\alpha$ is greater (because of the presence of more antenna chlorophylls) than that of PSII$_\beta$. Little is understood of the mechanism of membrane stacking into grana and the biochemical and physiological significance of this phenomenon although we know that the degree of stacking is influenced by the cation concentration and by the phosphorylation of light-harvesting proteins.

8.4 The electron transport sequences

The schemes given in Chapter 5 are certainly not final and will be modified as more is learnt about the reaction kinetics of individual components of the electron transfer chain and with the discovery of new components by fast scanning (pico and femto second) laser flash spectroscopy. The orientation of the individual components in the membrane and their movement across and along the membrane are also of importance and are the subject of considerable research. The role of the large pool of plastoquinones between photosystems II and I is intriguing.

8.5 Energy redistribution between the two photosystems

The Z scheme of photosynthesis envisages that for photosynthesis to proceed efficiently there must be equal input of light energy to the reaction centres of both photosystems. This is not a problem at saturating light intensities. But at low light intensities or with light of a limited spectral range an imbalance can occur and one of the photosystems could be preferentially excited over the other. However, oxygen evolution measurements and fluorescence data have shown that the steady state quantum yields for photosynthesis are constant and maximal over a wide spectral range where PSII absorption predominates. This suggests the existence of a mechanism in the photosynthetic apparatus for redistributing the energy absorbed by the light-harvesting pigments between the two photosystems so as to optimize the electron flow.

For example, if algae are illuminated with light below 680 nm, which preferentially excites PSII, part of the absorbed energy is transferred from the antenna chlorophylls of PSII to those of PSI; this transfer of excess energy from PSII to PSI is known as *spillover*. Spillover can be stopped by illumination of the algae with light above 680 nm which excites PSI. The redistribution of excitation energy by spillover from overexcited PSII (State 2) to PSI is referred to as the State 2 to State 1 transition. Conversely, the adaptation following over-excitation of PSI (State 1) and inhibition of spillover is referred to as the State 1 to State 2 transition. The phenomena of State 2–State 1 transitions have been observed in all types of oxygen-evolving photosynthetic organisms. Plants growing in shaded areas and algae living in ocean depths have adapted these transition mechanisms very efficiently for maximizing their photosynthetic capacity.

How does the photosynthetic apparatus adapt itself to the variations in light intensities? It has been known for some time that spillover is brought about by conformational changes of the chloroplast membrane. Biochemical and structural studies initiated within the last decade have provided a better understanding of the regulation of the State 1–State 2 transitions (see Allen, 1983 and Staehelin and Arntzen, 1984). It is now believed that these transitions are controlled by the redox state of the quinones in the plastoquinone pool (Fig. 5.2, p. 54).

When incident light energy begins to drive electrons from PSII at a faster rate than PSI can accept them there is an accumulation of reduced plastoquinone which in turn activates a membrane-bound protein kinase enzyme. The activated protein kinase catalyzes the phosphorylation of the light-harvesting chlorophyll proteins in the grana membranes where PSII predominates. The addition of negative (phosphate) charges to the pigment–protein complex reduces the membrane adhesion forces, unstacks the thylakoids, and promotes the lateral movement of the antenna pigments to adjacent stroma membranes where PSI predominates (Fig. 8.3). When PSI is overexcited plastoquinol is oxidized rapidly to plastoquinone, the protein kinase is deactivated, and the PSII pigments stay stacked in the grana. The phosphorylated proteins are dephosphorylated by phosphatases in both photosystems. Fluorescence, ^{32}P labelling and fractionation of membrane proteins, freeze-fracture electron microscopy, and other techniques support this hypothesis. However, this type of lateral migration of phosphorylated pigment proteins from grana to stroma regions cannot explain the spillover phenomenon observed in blue-green algae which do not possess chlorophyll *a/b* proteins as antenna pigments and are devoid of grana stacks.

8.6 Fluorescence as a probe for energy transfer in photosynthesis

It was mentioned in Chapter 4 that some of the energy quanta absorbed by chlorophyll are emitted as fluorescence in the red. The intensity of this fluorescence increases at low temperatures (77 K, liquid N_2) and so low temperature fluorescence spectroscopy has proved to be a valuable probe in elucidating many photochemical processes. The parameters usually measured are the wavelengths of fluorescence emission bands, the shape of fluorescence bands and fluorescence life time.

By careful analysis of low temperature fluorescence of algal cells, of isolated chloroplasts and of individual PSI and PSII particles the source of fluorescence emission bands at five different wavelengths has been tentatively identified. These emission bands are at 680 nm or F680 (from light-harvesting chlorophyll protein, LHCP), F685 (mainly from PSII antenna chlorophyll), F695 (PSII and PSI antenna chlorophyll), F720 (PSI antenna chlorophyll of algae) and F735 (PSI antenna chlorophyll of chloroplasts).

When dark-adapted chloroplasts are brightly illuminated their chlorophyll fluorescence rises immediately from an initial low intensity (F_0) to a maximum (F_m) and then gradually declines to a steady state fluorescence (F_s). There is good correlation between the kinetics of the fluorescence rise and the decay of oxidized PSII reaction centre chlorophyll, $F680^+$. However, fluorescence rise curves are complex and have been shown to consist of a variable fluorescence (F_v) contributed by the PSII reaction centre superimposed upon a relatively weaker constant background fluorescence (F_0) of PSI ($P700^+$) and of light-harvesting chlorophyll *a* molecules. Fluorescence emission and quenching are sensitive to a wide variety of photochemical events such as the redox state of donors and acceptors of the PSII reaction centre, proton translocation, degree of stacking and cation concentration of thylakoids, pH, CO_2 fixation rate, etc. However, notwithstanding these difficulties, the advantage of using a non-destructive technique in leaves to measure their photosynthetic capability is finding wide application in both crop and natural plant systems.

8.7 Mechanism of ATP formation

The relationship of the proton pump and membrane field gradient (Mitchell's chemiosmotic hypothesis) to the formation of ATP is being investigated using subchloroplast membrane particles, antibodies, algal mutants, inhibitors and uncouplers, coupling factors and ATPase enzymes, electron transport control parameters in whole and broken chloroplasts, and various other techniques. The sites of ATP formation are also not clearly defined. There are two 'sites' of ATP formation coupled to proton translocation associated with Photosystems I and II. The precise stoichiometry of photophosphorylation ($NADPH_2$: ATP) is still uncertain, i.e. is the $ATP/2e^-$ ratio 1.0, 1.33 or 2.0; the importance of an understanding of the correct ratio is relevant to the optimum number of quanta required to fix one molecule of CO_2. Significant progress has been made on understanding the structure of the ATPase – the 5 different subunits are shown in Fig. 5.2, p. 55. The properties of the subunits are becoming understood – membrane binding, inhibitor activity, proton exchange, phosphate binding, etc.

8.8   The photosystem II oxygen-evolving reaction

This key reaction of plant and algal photosynthesis is the main source of all oxygen on Earth but is still one of the unsolved mysteries in photosynthesis research. However, recent work has significantly helped to probe the mechanisms involved in splitting water to oxygen, protons, and electrons. Advances have been reported in the isolation and biochemical characterization of the polypeptides associated with the O_2-evolving complex, in the detection of Mn states by EPR (electron paramagnetic resonance) spectroscopy, and in correlating the sequences of the S-state cycle (§4.11, p. 49) with the transfer of electrons from the reaction centre. However, the isolation of a functional O_2-evolving complex from the photosynthetic membrane has still not yet been achieved. We do not understand how the various polypeptides associated with the complex are arranged in the membrane, what chemical intermediates are formed during water oxidation, how many Mn atoms are essential for catalysis, what is the role of cytochrome b_{559} in O_2 evolution, and whether chloride and calcium ions are required for the functioning of the O_2-evolving complex.

8.9   Role of light in the regulation of photosynthesis

The activity of many enzymes of the CO_2 fixation pathway are modulated by light. These include NADP-glyceraldehyde-3-phosphate dehydrogenase, fructose 1,6-bisphosphatase, sedoheptulose 1,7-bisphosphatase, RuBisCO, phosphoribulokinase, pyruvate phosphate dikinase and malate dehydrogenase. Several mechanisms have been proposed to explain the light activation of these enzymes such as change in stromal pH, increase in stromal Mg^{2+} concentration, change in redox state, and the mediation of protein factors and 'effectors'. Only the reduced forms (with vicinal SH groups) of enzymes such as fructose bisphosphatase, sedoheptulose bisphosphate, and glyceraldehyde 3-phosphate dehydrogenase are active in the Calvin cycle. Their activity therefore is governed by the competition for electrons between PSI electron acceptors (NADP and O_2) and the enzymes themselves.

Illumination of chloroplasts increases the stromal pH from 7 to 8 or higher (due to uptake of protons from the stroma to the thylakoids) and increases Mg^{2+} concentration (due to a counter movement of Mg^{2+} from the thylakoids to the stroma). Both increases enhance the catalytic activities of the above mentioned enzymes of the Calvin pathway. The activation of RuBisCO is affected also by other factors such as concentrations of sugar phosphates, CO_2, and reductants generated by photosynthetic electron transport. Fructose 2,6-bisphosphate, a modulator of the enzyme fructose 1,6-bisphosphatase, is another key metabolite exerting a regulatory effect on metabolic carbon interchange between the chloroplast and cytosol.

Light modulation of the C_4 pathway enzyme pyruvate dikinase, which catalyses the reaction:

$$\text{Pyruvate} + \text{Pi} + \text{ATP} \rightleftharpoons \text{phosphoenol pyruvate} + \text{AMP} + \text{pyrophosphate}$$

involves the interconversion between an active, non-phosphorylated form and an inactive, phosphorylated form of the enzyme. The level of ascorbate in the stroma is high and the redox state of glutathione and its reductase may also play a role in the regulation of chloroplast metabolism. Two chloroplast proteins thioredoxin and ferredoxin-thioredoxin reductase have been shown as components of a ferredoxin-linked regulatory mechanism of at least four reductive pentose phosphate cycle enzymes.

8.10 Exchange of ions and metabolites through the chloroplast envelope

Electron micrographs of chloroplasts isolated from higher plants show that they are bound by two distinct membranes which are together called the chloroplast envelope (Fig. 3.2b, p. 20). These envelopes are the site of synthesis of a number of membrane constituents and are involved in the transport of many metabolites and have thus attracted considerable attention. The envelopes can be separated from intact chloroplasts (or protoplasts) by gentle osmotic shock followed by centrifugation in a discontinuous sucrose gradient – the lighter envelopes sediment above the heavier thylakoids in the gradient. Techniques have also been developed to separate the inner from the outer membranes of the isolated envelopes. Purified envelopes are yellow in colour (due to carotenoids) and are free of chlorophyll, cytochrome b and oxidoreductase enzymes. The envelope has been shown to be the site of synthesis and assembly of membrane galactolipids, α-tocopherol and carotenoids. The polypeptide composition of the envelope is different from that of the thylakoids and of the stroma. Polyacrylamide gel electrophoresis of envelope proteins released by detergent treatment shows 75 different polypeptides; some of these are involved in regulating the exchange of metabolites through the envelope and others in maintaining the structure of the membrane.

The chloroplast envelope functions both as a barrier separating the stroma from the cytosol and as a bridge allowing the transport of specific metabolites between these two compartments. The permeability of various molecules through the chloroplast envelope has been studied (a) by feeding $^{14}CO_2$ to leaves and following the distribution of radioactively labelled metabolites in the chloroplast and cytosol, and (b) by adding metabolites to a suspension of *intact* chloroplasts and measuring the kinetics of specific photosynthetic reactions. Such studies have shown that the outer envelope membrane acts as a molecular sieve and is permeable to all molecules up to a Mr of 10 kD. The inner membrane (osmotic barrier) is almost impermeable to charged species such as ferredoxin, NAD(P), NAD(P)H, acetyl CoA, pyrophosphate, many sugar phosphates, small cations (Na^+, Mg^{2+}), and to sucrose and sorbitol. Uncharged molecules such as CO_2, O_2, acetic acid, pyruvic acid, glycerol, etc. can diffuse through the inner membrane. Passage of charged species such as Pi and triose phosphates, ATP, and dicarboxylate anions (malate, aspartate, oxaloacetate etc.) through the inner membrane occurs only in the presence of specific carriers present in the envelope called *translocators*. Of these translocators, the proteinaceous factor (phosphate or C_3 translocator) which allows the transport of triose phosphates from the stroma to the cytosol in exchange for the import of Pi into the stroma has been studied in most detail. The importance of this counter exchange in sucrose synthesis has been stressed earlier (§6.5). The chloroplast envelope also mediates in the export of fatty acids synthesized in the chloroplast to other organelles (for lipid synthesis) and the import of proteins synthesized in the cytoplasm into the chloroplasts (§8.2).

What is the exact nature of the polar lipid interactions with plastid membrane proteins? How are the membrane components which are synthesized in the envelope subsequently transferred and assembled in the thylakoids? Where do the polyunsaturated fatty acids in the chloroplasts originate? What are the specific functions of various envelope proteins? These are some of the questions for which answers are still needed.

8.11 Photoinhibition

The photosynthetic efficiency of many plants is decreased when they are subjected to stress conditions such as high light intensity, low temperature, low CO_2 environment, water limitation, etc. Photoinhibition is defined as the decrease in photosynthetic capacity induced by the exposure of photosynthetic tissues to high fluxes of photosynthetically active radiation (400–700 nm). Not included in this definition of photoinhibition are inhibition of photosynthesis caused by exposure to UV radiation, and photooxidation (which usually follows photoinhibition) leading to the bleaching of photosynthetic pigments. In most cases photoinhibition is reversible – normal photosynthetic rates are restored, after a lag period, when the organism is exposed to less intense irradiance. Ultrastructure studies using the electron microscope reveal no permanent damage to the thylakoid membranes during photoinhibition. The chloroplast is usually able to regenerate any damaged components. Photoinhibition in plants has been observed under three types of conditions: when a plant is exposed to higher irradiance than that under which it has been grown, when the plant is subjected to conditions which decreases its rate of carbon metabolism, and when some species are exposed to chilling temperatures (10°C and below) even under normal irradiance.

The extent of photoinhibition can readily be determined by measuring photosynthetic electron transport as oxygen exchange or fluorescence (induction and decay) which reflect electron transfer around the two photosystems. Photoinhibition results in a decrease of quantum yield which is related to the photosynthetic photon flux density (PPFD) to which the plants are exposed. In isolated chloroplasts photoinhibition is much more evident in PSII activities than in PSI. Measurement of photosynthetic electron transport through various segments of chloroplast membranes isolated from photoinhibited algae has shown a definite correlation between photoinhibition and the inactivation of a 32 kD polypeptide which binds Q_B, a PSII electron-mediator (Fig. 5.2). Other studies have indicated a decrease in the content of one of two polypeptides around the reaction centre of PSII as a result of photoinhibition. Photoinhibition in PSI may be due to an accumulation of electrons over and above those which can be transferred to NADP.

Comparative analysis of some leaf enzymes from chilling-sensitive and chilling-resistant plants have indicated a lower content of superoxide dismutase and catalase, two enzymes involved in oxygen metabolism, in the chilling-sensitive strains. Thus, there are many factors already known and many still to be discovered which contribute to photoinhibition. The identification of these factors and development of techniques to minimize photoinhibition both in the field (whole plants and algae) and in the laboratory (isolated chloroplasts) are currently active areas of fundamental photosynthesis research which may lead to practical applications.

8.12 Whole plant studies and bioproductivity

The efficiency of photosynthesis of the whole plant is crucial to agriculture, forestry, ecology, etc. when it comes to analysing productivity for food and fuels and many other product uses. The quantity and quality of incident light (photosynthetic photon flux density, PPFD), temperature and water stresses, availability and utilization of mineral nutrients, photorespiratory losses, presence of pollutants in the atmosphere (NO_2, SO_2, O_3) and in the soil (heavy metals), etc., are some of the factors which affect plant productivity.

The amount of incident radiation that can be captured by the plant will depend on its canopy and leaf structure. Photosynthesis is usually saturated at moderate light intensity – photosynthesis will function effectively at a PPFD of 200 μmol m^{-2} s^{-1}, about one-tenth of the PPFD encountered by plants when the sun is directly overhead on a cloudless day. In addition many plants have the capacity to adapt to different light intensities during growth. Information about the relative content of the two photosystems (number of antenna and reaction centre chlorophylls, grana and stroma lamellae, etc.) in leaves can be obtained by ultrastructure analysis and by absorption spectroscopy of the chloroplasts isolated from the leaves. These analyses show that the PSII/PSI ratios are higher in shade-adapted species which receive far-red enriched

($\lambda \geqslant 700$ nm) light, than in sun-adapted species which receive far-red depleted ($\lambda < 700$ nm) illumination. Thus the specific aspects of chloroplast structure, composition and function are determined by the light conditions available for plant growth.

Temperature is another factor determining plant productivity – all photosynthetic reactions with the exception of the primary photoacts are thermochemical. Quantum yields are the same at 30°C for both C_3 (optimal temperature) and C_4 plants; for C_4 plants (growing in the tropics) the optimum is usually at a higher temperature.

Availability of fixed nitrogen is one of the important factors contributing to plant productivity. It is taken up from the soil as NH_4^+ or NO_3^-, and in leguminous species by direct fixation of atmospheric N_2 by soil bacteria (Rhizobia) symbiotically associated with root nodules. Many species of blue-green algae (cyanobacteria) also fix atmospheric N_2. One such N_2-fixing blue-green alga which has application in agriculture is *Anabaena azollae* (Fig. 3.4c, p. 22) which occurs in natural populations as a symbiont with an aquatic fern, *Azolla*. The *Azolla–Anabaena* symbiont is widely used as a source of nitrogen fertilizer in rice fields throughout Asia. *Anabaena azollae* can be separated from the host fern by mechanical means and then grown in the laboratory in the presence of synthetic foam pieces (polyvinyl or polyurethane). Under these conditions the blue-green algae adhere in the foam pores (Fig. 8.4) and are immobilized. Such foam-immobilized *A. azollae* maintain their photosynthetic activities for very long periods – nitrogen fixation activity being observed for several months.

The importance in compiling productivity data from simultaneous measurements of all the integrated photosynthetic activities of whole leaves (or canopies), rather than individual reactions of cells or chloroplasts is being recognized now. Instruments have been devised for studying photosynthesis in illuminated leaves with the simultaneous determination of O_2 and CO_2 exchanges and fluorescence emission (which is indicative of the energized state of the chloroplast membrane). Walker and others have used this technique to study the effect of added nutrients (phosphate, nitrate, etc.) and sugars (mannose) on whole leaf photosynthesis. The technique is also useful for rapidly assaying the comparative photosynthetic activities of wild and mutant strains of plants. Such nondestructive probes will be useful tools for crop physiologists and plant breeders.

8.13 Photosynthesis and the 'greenhouse' effect

The global annual extraction of fossil fuels (coal, oil, etc.) now exceeds 8 billion (10^9) tonnes. The combustion of these fossil fuels and the burning of biomass (wood, straw, etc.) generates large quantities of particulate matter (fly ash), 'greenhouse' gases (oxides of carbon, sulphur and nitrogen, and including ozone and chlorofluorocarbons) and heat. Human activities now release into the troposphere an estimated 2.5 billion tonnes of particulate matter (fly ash and aerosol), 180 million tonnes of sulphur dioxide and 40 million tonnes of

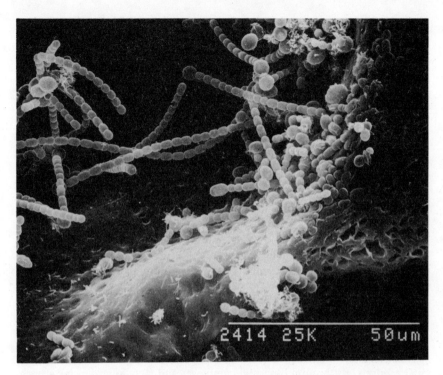

Fig. 8.4 Scanning electron micrograph of the cyanobacterium *Anabaena azollae* immobilized in polyvinyl foam pieces. (Courtesy of D. J. Shi and G. Morgan, King's College, London.)

oxides of nitrogen. The carbon dioxide content of the atmosphere has gone up from 270 to 345 parts per million (ppm) in the span of a century and is estimated to reach 600 ppm in 50 years' time if we continue to generate CO_2 on past trends. Increasing concentrations of greenhouse gases are expected to cause a significant warming of the global climate in the next century – the effect being felt more at the high latitudes than in the tropics. Computer simulations of model ecosystems forecast temperature rises of 1.5 to 4.5°C for every doubling of CO_2 concentration. These temperature changes and accompanying variations in rainfall will occur unevenly over the global surface affecting some countries and regions more than others. If these predicted climatic changes come true then they would have a profound effect on global ecosystems and agriculture. Naturally it has created great concern among scientists, politicians and others.

Is there any reason for alarm? Even though the indiscriminate release of particulate matter is detrimental to the environment one should be aware of the beneficial effects of CO_2 especially, and NO_2 to a lesser extent, on photosynthetic productivity. Oxides of nitrogen are oxidized to nitrates and increase soil fertility. Net photosynthesis by plants is promoted by higher levels of ambient carbon dioxide. Increased carbon dioxide levels also result in better water and fertilizer use by plants and lower the photorespiration rate (in C_3 plants). However, temperature increases may cause changes in the rainfall pattern especially in the tropics. Remote sensing (using satellite) studies are now being undertaken to monitor biomass production in plant canopies (from measurements of scattered non-photosynthetically active radiation, 700–1100 nm) and estimation of carbon dioxide concentration. However, as yet we understand very little about the complex interrelations which exist in plants and ecosystems when CO_2 and temperature levels are increased over the long term. More physiological, agricultural and ecological research is urgently required if we are to avert any possible negative consequences of the greenhouse effect.

8.14 Mimicking photosynthesis

Photochemists and photobiologists are actively searching for synthetic systems which will split water using solar energy. The advantage of these artificial systems over natural photosynthesis is that they might be optimized for maximum photosynthetic efficiency since they are not limited by the inherent physiological characteristics and requirements of whole plants. To be effective such artificial devices mimicking biological processes (biomimetic systems) should be able to carry out the essential steps in natural photosynthesis viz. light absorption and energy migration, charge separation, and electron transfer and catalysis. Ruthinium bipyridyl complexes and metalloporphyrins absorb light of visible wavelengths, have high extinction coefficients, have suitable redox potential differences between the ground and excited states, and good stability in light, and therefore are generally used as photosensitizers. Incorporation of the photosensitizer and electron donor in micelles or emulsions enhances the stability of the charged species generated after photon capture.

Light-induced electron transport has been demonstrated in synthetic membranes made up of liposomes embedded with pigments, proteins and other catalysts. Homogeneous or heterogeneous systems consisting of an electron donor, photosensitizer, electron mediator and hydrogen activation catalyst (platinum or hydrogenase) are now available for the photoproduction of hydrogen via a PSI-type of electron transport. Photooxidation of water can be achieved using ultrafine semi-conductor particles of TiO_2 in the presence of RuO_2. Research is continuing on the synthesis of a manganese complex which would catalyse the splitting of water with O_2 evolution thus mimicking the water-oxidizing complex of PSII.

One of the difficulties encountered in photoactive biomimetic systems is the high back reaction rates, i.e. the rapid recombination of the photogenerated charged species before electron transfer to an acceptor can be accomplished. Some success towards overcoming this problem has been achieved by the recent synthesis of a tripartite molecule which mimics the properties of a natural photosynthetic reaction centre. This so-called molecular triad consists of a porphyrin derivative (chromophore) sandwiched between a carotenoid (e^- donor) and a quinone (e^- acceptor) via covalent linkages. Laser flash excitation of this molecule generated within 100 picoseconds a charge–transfer complex (from the singlet state of the excited porphyrin) with a life time in the microsecond time scale – the back reaction was 100 times slower than the forward reaction. The quantum yield for the formation of the complex was about 25% and the energy stored in the complex was more than 1 eV above the ground state. The similarity of the synthetic molecule to the photosynthetic reaction centre was extended further by the observation that 10% of the light energy absorbed by the carotenoid was transferred to the porphyrin.

Thus research is progressing in diverse directions towards the construction of an artificial photosynthetic apparatus. However, even though considerable progress has been made in recent years in devising systems for water photolysis, ATP synthesis from the coupling of photosynthetic electron transfer in such systems to the vectorial flow of electrons and protons through membranes still remains as a challenging problem.

9
Laboratory Experiments

9.1 Reference books for experiments

Coombs, J., Hall, D.O., Long, S.P. and Scurlock, J.M.O. (eds) (1985). *Techniques in Bioproductivity and Photosynthesis*, 2nd edition. Pergamon Press, Oxford.

Hall, D.O. and Hawkins, S.E. (eds) (1975). *Laboratory Manual of Cell Biology*. The English University Press, Hodder and Stoughton, London.

Machlis, L. and Torrey, J.G. (1959). *Plants in Action*. W.H. Freeman, San Francisco.

Meidner, H. (1984). *Class Experiments in Plant Physiology*. George Allen and Unwin, London.

Walker, D.A. (1986). *Use of the Oxygen Electrode and Fluorescence Probes in Simple Measurements of Photosynthesis*. Oxygraphics Ltd. University of Sheffield Print Unit.

See also

Hipkins, M.F. and Baker, N.R. (1986) Photosynthesis energy transduction — a practical approach. IRL Press Oxford.

Marshall, B. and Woodward F.I. (eds.) (1985) Instrumentation for environmental physiology. Cambridge University Press.

Packer, L. and Douce, R. (eds) (1987) *Methods in Enzymology*. Vol. 148. Plant Cell Membranes. Academic Press, New York.

San Pietro, A. (ed.) (1980). *Methods of Enzymology*, Vol. 69, part C. Photosynthesis and N_2 fixation. Academic Press, New York.

9.2 Photosynthesis in whole plants

(a) Measurement of photosynthetic gas exchange using infra-red gas analyser (IRGA) (Coombs *et al.*, p. 90; Meidner, p. 102).
(b) $^{14}CO_2$ incorporation into leaves (Coombs *et al.*, pp. 75 and 150; Meidner, p. 109).
(c) Measurement of O_2 evolution and chlorophyll fluorescence in leaves (Coombs *et al.*, p. 95; Meidner, p. 105).
(d) Measurement of sucrose and starch in leaves (Coombs *et al.*, p. 125).
(e) Measurements of CO_2 compensation point and photorespiration (Coombs *et al.*, p. 151; Meidner, p. 104).
(f) Measurement of light compensation point (Meidner, p. 103).
(g) Fluorescence measurements (Walker).

9.3 Photosynthesis in Elodea and algae

(a) O_2 evolution in *Elodea*; influence of light intensity and temperature (Machlis and Torrey, p. 132).
(b) Photosynthetic action spectra of *Scenedesmus* or *Chlorella* measured as O_2 evolution (Meidner, p. 107).

(c) Batch culture techniques for algae (Coombs *et al.*, p. 189).

9.4 Preparation of protoplasts and chloroplasts

(a) Preparation of protoplasts from leaves of spinach and sunflower (Coombs *et al.*, p. 120).
(b) Preparation of chloroplasts (Coombs *et al.*, p. 120; see also Chapter 3 in this book).

9.5 Separation and estimation of photosynthetic pigments

(a) Pigment extraction; acetone extraction and separation into petroleum ether and methanol (Machlis and Torrey, p. 136).
(b) Acetone extraction of pigments from leaves, algae, and cyanobacteria and separation by thin layer chromatography (Meidner, p. 111).
(c) Separation of pigments of photosynthetic membranes by HPLC (Lichtenthaler, H.K. in Packer and Douce, p. 350).

9.6 Photosynthetic electron transport using oxygen electrode and/or spectrophotometer

(a) Whole chain electron transport (PSII and PSI) from H_2O to ferricyanide measured as O_2 evolution in the absence and presence of the uncoupler, ammonium chloride: determination of chloroplast intactness (Coombs *et al.*, p. 126; Hall and Hawkins, p. 151).
(b) Determination of whole chain electron transport from water to methyl viologen measured as O_2 uptake in the presence of sodium azide (catalase inhibitor) or as O_2 evolution in the absence of azide (Coombs *et al.*, p. 137; Hall and Hawkins, p. 151).
(c) Determination of Hill reaction in chloroplasts by reduction of the dye dichlorophenol indophenol (DCPIP) measured spectrophotometrically (Hall and Hawkins, p. 151; Machlis and Torrey, p. 141; Meidner, p. 110).
(d) Determination of PSII electron transport from H_2O to the dye p-phenylenediamine in the presence of ferricyanide measured as O_2 evolution (Coombs *et al.*, p. 137).

9.7 Proton flux and photophosphorylation

(a) Measurement of transmembrane proton flux in illuminated 'broken' chloroplasts using a pH electrode (Coombs *et al.*, p. 134).
(b) Determination of ATP formation from orthophosphate and ADP by measuring the rate of alkalination of chloroplasts in light using a pH electrode (Coombs *et al.*, p. 135).
(c) Incorporation of ^{32}P-labelled orthophosphate into ATP by chloroplasts in light measured by radioisotope tracer technique (Hall and Hawkins, p. 36).

Appendix

Chemical Names

The names of many of the chemical substances discussed in the text have undergone changes during the last few years; the following list may be useful.

Old name	New Name
Acetaldehyde	Ethanal
Acetic acid	Ethanoic acid (old name still acceptable)
Acetylene	Ethyne
Citric acid	2-hydroxypropane-1,2,3-tricarboxylic acid
Ethyl alcohol	Ethanol
Ethylene	Ethene
Fumaric acid	*Trans*-butanedioic acid
Glutamic acid	2-aminopentanedioic acid
Iso-citric acid	1-hydroxypropane-1,2,3-tricarboxylic acid
α-ketoglutaric acid	1-oxobutanedioic acid
Malic acid	2-hydroxybutanedioic acid
Malonic acid	propanedioic acid
Oxaloacetic acid	2-oxobutanedioic acid
Oxalosuccinic acid	1-oxopropane-1,2,3-tricarboxylic acid
Pyruvic acid	2-oxopropanoic acid
Succinic acid	Butanedioic acid

Further Reading

Non-specialist books

Edwards, G.E. and Walker, D.A. (1983). C_3, C_4: *Mechanisms and Cellular and Environmental Regulation of Photosynthesis*. Blackwell Scientific Publications, Oxford.

Foyer, C.H. (1984). *Photosynthesis*. John Wiley, New York.

Halliwell, B. (1984). *Chloroplast Metabolism*. Clarendon Press, Oxford.

Hoober, J.K. (1984). *Chloroplasts*. Plenum Press, London.

Lange, D.O., Nobel, P.S., Osmond, C.B. and Ziegler, H. (eds) (1981–83). *Physiological Plant Ecology, I–IV. Encyclopedia of Plant Physiology*, Vols 12A to 12D. Springer-Verlag, Berlin.

Lawlor, D.W. (1987) Photosynthesis: metabolism, control and physiology. Longman, Harlow, England.

Tribe, M. and Whittaker, P. (1981). *Chloroplasts and Mitochondria*. Studies in Biology No. 31. Edward Arnold, London.

Whatley, J.M. and Whatley, F.R. (1980). *Light and Plant Life*. Studies in Biology No. 124. Edward Arnold, London.

'Scientific American' offprints

Bjorkman, O. and Berry, J. (1973). *High efficiency photosynthesis*, **229 (4)**, 80–93.

Govindjee and Govindjee, R. (1974). *The Absorption of Light in Photosynthesis*, **231(6)**, 68–82.

Hinkle, P.C. and McCarty, R.E. (1978). *How cells make ATP*, **238 (3)**, 104–23.

Miller, K.R. (1979). *The photosynthetic membrane*, **241 (4)**, 100–13.

Moses, P.B. and Chua, N-H (1988) *Light Switch for Plant Genes*, **258 (4)**, 64–69.

Youvan, D.C. and Marrs, B. (1987) *Molecular Mechanisms of Photosynthesis*, **256 (b)**, 42–48.

'Trends in Biochemical Sciences' offprints

Allen, J.F. (1983). Protein phosphorylation-carburettor of photosynthesis?, **8**, 369–73.

Anderson, J.M. and Andersson, B. (1982). The architecture of photosynthetic membranes, **7**, 288–92.

Blankenship, R.E. and Prince, R.E. (1985). Excited state redox potentials and the Z scheme of photosynthesis, **10**, 382–83.

Burnell, J.N. and Hatch, M.D. (1985.) Light–dark modulation of leaf pyruvate, Pi dikinase, **10**, 288–91.

Carillo, N. and Vallejos, R.H. (1983). The light-dependent modulation of photosynthetic electron transport, **8**, 52–6.

Cramer, W.A., Widger, W.R., Herrmann, R.G. and Trebst, A. (1985). Topography and function of the thylakoid membrane proteins, **10**, 125–9.

Deisenhoffer, J., Michel, H. and Huber, R. (1985). The structural basis of photosynthetic light reactions in bacteria, **10**, 243–8.

Ellis, R.J. (1979). The most abundant protein in the world, **4**, 241–4.

Flügge, I.U. and Heldt, H.W. (1984). The phosphate-triose phosphate-phosphoglycerate translocator of the chloroplast, **9**, 530–3.

Foyer, C.H. and Hall, D.O. (1980). Oxygen metabolism in the active chloroplast, **5**, 188–91.

Hatch, M.D. (1977). C_4 pathway of photosynthesis: mechanism and physiological function, **2**, 199–202.

Heber, U. and Walker, D.A. (1979). The chloroplast envelope – barrier or bridge?, **4**, 252–6.

Miller, K.R. and Lyon, M.K. (1983). Do we really know why chloroplast membranes stack?, **10**, 219–22.

Murata, N. and Miyao, M. (1985). Extrinsic membrane proteins in the photosynthetic oxygen-evolving complex, **10**, 122–4.

Nugent, J.H.A. (1984). Photosynthetic electron transport in plants and bacteria, **9**, 354–7.

Sommerville, C.R. and Ogren, W.L. (1982). Genetic modification of photorespiration, **7**, 171–4.

Thornber, J.P. and Markwell, J.P. (1981). Photosynthetic pigment–protein complexes in plant and bacterial membranes, **6**, 122–5.

Reviews and articles in 'Photosynthesis Research'

Blankenship, R.E. (1985). Electron transport in green photosynthetic bacteria, **6**, 317–33.

Leegood, R.C. (1985). Regulation of photosynthetic CO_2 pathway enzymes by light and other factors, **6**, 247–59.

Merchant, S. and Selman, B.R. (1985). Photosynthetic ATPases: purification properties, subunit isolation and function, **6**, 3–31.

Renger, G. and Govindjee (1985). The mechanism of photosynthetic water oxidation, **6**, 33–55.

Rutherford, A.W. and Heathcote, P. (1985). Primary photochemistry in photosystem I, **6**, 295–316.

Schreiber, U. Schliwa, U. and Bilger, W. (1986). Continuous recording of photochemical and non-photochemical chlorophyll fluorescence quenching with a new type of modulation fluorometer, **10**, 51–62.

Van Gorkow, H.J. (1985). Electron transfer in photosystem II, **6**, 97–112.

Williams W.P. and Allen J.F. (1987). State 1/State 2 changes in higher plants and algae, **13**, 19–45.

Reviews in Photochemistry and Photobiology

Bose, S.C. (1982). Chlorophyll fluorescence in green plants and energy transfer pathways in photosynthesis, **36**, 725–31.

Dismukes, G.C. (1986). The metal centres of the photosynthetic oxygen-evolving complex, **43**, 99–115.

Govindjee, Kambara, T. and Coleman, W. (1985). The electron donor side of Photosystem II: the oxygen-evolving complex, **42**, 187–210.

Haworth, P., Kyle, D.J., Horton, P. and Arntzen, C.J. (1982). Chloroplast membrane protein phosphorylation, **36**, 743–48.

Percomaro, V.L. (1988) Structural proposals for the manganese centres of the oxygen evolving complex. **48**, 249–64.

Kyle, D.J. (1985). The 32000 dalton Q_B protein of photosystem II, **41**, 107–116.

Percomaro, V.L. (1988) Structural proposals for the manganese centres of the oxygen evolving complex. **48**, 249–64.

Satoh, K.C. (1985). Protein pigments and photosystem II reaction centre, **42**, 845–53.

Zuber, H. (1985). Structure and function of light-harvesting complexes and their polypeptides, **42**, 821–44.

Annual Review of Plant Physiology

Anderson, J.M. (1986). Photoregulation of the composition, function and structure of thylakoid membranes, **37**, 93–136.

Andreasson, L.E. and Vanngard, T. (1988) Electron Transport in Photosystems I and II, **39**, 379–411.

Badger, M.R. (1985). Photosynthetic oxygen exchange, **36**, 27–53.

Barber, J. (1982). Influence of surface charges on thylakoid structure and function, **33**, 261–95.

Cogdell, R.J. (1983). Photosynthetic reaction centres, **34**, 21–45.

Fork, D.C. and Satoh, K. (1986). The control by state transitions of the distribution of excitation energy in photosynthesis, **37**, 335–61.

Glazer, A.N. and Melis A. (1987). Photochemical reaction centres: structure, organization and function, **38**, 11–45.

Haehnel, W. (1985). Photosynthetic electron transport in higher plants, **35**, 659–93.

Mullet, J.E. (1988) Chloroplast development and gene expression, **39**, 475–502.

Ogren, W.L. (1984). Photorespiration: pathways, regulation and modification, **35**, 415–42.

Powles, S.B. (1984). Photoinhibition of photosynthesis induced by visible light, **35**, 14–44.

Somerville, C.R. (1986). Analysis of photosynthesis with mutants of higher plants and algae, **37**, 467–507.

Strotmann, H. and Bickel-Sandkotter, S. (1984). Structure, function and regulation of chloroplast ATPase, **35**, 97–120.

Ting, I. (1985). Crassulacean acid metabolism, 36, 216–34.

Tobin, E.M. and Silverthorne, J. (1985). Light regulation of gene expression in higher plants, **36**, 569–93.

Whitfield, P.W. and Bottomley, W. (1983). Organization and structure of chloroplast genes, **34**, 279–310.

Woodrow, I.E. and Berry, J.A. (1988) Enzymatic regulation of photosynthetic CO_2 fixation in C_3 plants, **39**, 533–94.

More specialized books and articles

Allen, J.F. (1983). Regulation of photosynthetic phosphorylation. *Critical Reviews in Plant Science*, **1**, 1–22, CRC Press, Boca Ratan.

Amesz, J. (ed). New Comprehensive Biochemistry Vol 15 *Photosynthesis*. Elsevier, Amsterdam.

Barber, J. (ed.) (1976–1987). *Topics in Photosynthesis*. Vols 1–9. Elsevier, Amsterdam.

Bassham, J.A. (1977). Increasing crop production through more controlled photosynthesis. *Science*, **1971**, 630–8.

Beadle, C.R., Long, S.P. Imbamba, S.K., Hall, D.O., and Olembo, R.J. (1985). *Photosynthesis in Relation to Plant Production in Terrestrial Environment*. Tycooly Publications Limited, Oxford.

Biggins, J. (ed.) (1986). *Proceedings of the VIIth International Congress on Photosynthesis*. Martinus Nighoff/Dr. W. Junk Pub., The Hague.

Bolton, J.R. and Hall, D.O. (1979). Photochemical conversion and storage of solar energy. *Annual Review of Energy*, **4**, 353–401.

Boyer, J.S. (1982). Plant productivity and environment. *Science*, **218**, 443–8.

Carlson, P. (ed.) (1980). *The Biology of Crop Productivity*. Academic Press, New York.

Clayton, R.K. and Sistrom, W.R. (eds) (1979). *The Photosynthetic Bacteria*. Plenum Press, New York.

Clayton, R.K. (1980). *Photosynthesis, Physical Mechanisms and Chemical Patterns*. Cambridge University Press, Cambridge.

Coombs, J., Hall, D.O., and Chartier, P. (eds) (1983). *Plants as Solar Collectors*. D. Reidel Publishing Company, Dordrecht.

Ellis, R.J. (ed.) (1984). *Chloroplast Biogenesis*. Cambridge University Press, Cambridge.

Fuller, K.W. and Gallow, J.R. (eds) (1985). *Plant Products and the New Technology*. Clarendon Press, Oxford.

Govindjee (ed.) (1982). *Photosynthesis*. Vols 1 and 2. Academic Press, New York.

Govindjee, Amesz, J. and Fork, K. (eds) (1986). Light emission by plants and bacteria. Academic Press, New York.

Hall, D.O. (1979). Solar energy through biology – past, present and future. *Solar Energy*, **22**, 307–23.

Hall, J.L. and Moore, A.L. (eds) (1983). *Isolation of Membranes and Organelles from Plant Cells*. Academic Press, London.

Hatch, M.D. and Boardman, N.K. (eds) (1981). *The Biochemistry of Plants*, Vol. 8, Photosynthesis. Academic Press, New York.

Hill, R. (1965). The Biochemist's green mansion: the photosynthetic electron transport chain in plants, in *Essays in Biochemistry*, Vol. 1. Academic Press, London.

Homann, P.H. (1988) Explorations in the 'Inner sanctum of the Photosynthetic Process', the water oxidising system. *Plant Physiol.* **88**, 1–5.

Kirk, J.T.O. and Tilney-Bassett, R.A.G. (1978). *The Plastids*, 2nd edition. Elsevier, Amsterdam.

Magnien, E. and De Nettancourt, D. (eds) (1985). *Genetic Engineering of Plants and Microorganisms Important for Agriculture*. Martinus Nijhoff/Dr. W. Junk Publishers, The Hague.

Nobel, P.S. (1983). *Biophysical Plant Physiology and Ecology*. W.H. Freeman, San Francisco.

Rao, K.K., Cammack, R. and Hall, D.O. (1985). Evolution of Light Energy Conversion. In *Evolution of Prokaryotes*, K.H. Schleifer and E. Stackebrant (eds). Academic Press, London. pp. 143–73.

Rao, K.K. and Hall, D.O. (1982). Photorespiration. *Journal of Biological Education*, **16(3)**, 167–72.

San Pietro, A. (ed.) (1980). *Methods in Enzymology*, Vol. 69, Part C, Photosynthesis and Nitrogen Fixation. Academic Press, New York.

Staehelin, L. and Arntzen, C.J. (1984). Regulation of chloroplast membrane function; protein phosphorylation changes the spatial organization of membrane components. *Cell Biology*, **97**, 1327–37.

Staehelin, L.S. and Arntzen, C.J. (eds) (1986). *Photosynthesis III. Encyclopedia of Plant Physiology*, Vol. 19. Springer-Verlag, Berlin

Stumpf, P.K. and Conn, E.E. (eds) (1980). *The Biochemistry of Plants*, Vol. 1, *The Plant Cell*, ed. N.E. Tolbert. (1981).

Sybesma, C. (ed.) (1984). *Advances in Photosynthesis Research*, Vol. 1–5. Martinus Nijhoff/Dr. W. Junk Publishers, The Hague.

Wortman, S. (1980). World food and nutrition: the scientific and technological base. *Science*, **209**, 157–64.

Zaborsky, O.R. (ed.) (1982). *Handbook of Biosolar Resources*, Vols 1 and 2. CRC Press, Boca Ratan.

Index